# International Series in Operations Research & Management Science

Volume 284

More information about this series at http://www.springer.com/series/6161

Roger Z. Ríos-Mercado

Editor

# Optimal Districting and Territory Design

 Springer

*Editor*
Roger Z. Ríos-Mercado
Graduate Program in Systems Engineering
Universidad Autónoma de Nuevo León
San Nicolás de los Garza
Nuevo León, Mexico

ISSN 0884-8289          ISSN 2214-7934   (electronic)
International Series in Operations Research & Management Science
ISBN 978-3-030-34314-9          ISBN 978-3-030-34312-5   (eBook)
https://doi.org/10.1007/978-3-030-34312-5

This Springer imprint is published by the registered company Springer Nature Switzerland AG.
The registered company address is: Gewerbestrasse 11, 6330 Cham, Switzerland

*To Biali, Shaní, and Vandari.*

# Preface

The purpose of this book is to collect the most recent advances from renowned scholars in the field of districting, territory design, and zone design. This book provides knowledge and insights on recent advances in problems where districting decisions are considered. The aim is to present contributions on models, theory, algorithms (exact or heuristic), and applications that would bring an update on the state of the art of districting systems. The book also captures the diverse nature of districting applications as the chapters are written by leading experts on political, sales, location/routing, and healthcare applications, to name a few. The target audience of this book are professionals, researchers, and graduate students from diverse fields such as operations research, management science, computer science, discrete mathematics, and regional science.

Each chapter was written by leading experts on some area of districting and was peer reviewed by two or more anonymous independent reviewers to ensure a high-quality volume. I am very grateful to the authors of the chapters in this book for agreeing to participate and collaborate in this project with great dedication and enthusiasm. It was a pleasure working with all of them along the process. I also would like to thank all our anonymous reviewers for their critical remarks and timely reports that helped improve the quality of each chapter. Their contribution to significantly improve the quality of this volume is very much appreciated. Unfortunately, their names cannot be listed due to the blind nature of the review process.

I also would like to thank the editorial team at Springer, particularly Matthew Amboy, former Senior Editor, Business & Economics: Operations Research & Management Information Systems, Neil Levine, Publishing Editor, and Faith Su, Assistant Editor, for providing guidance and support throughout the entire editorial process.

San Nicolás de los Garza, Mexico
August 2019

Roger Z. Ríos-Mercado

# Contents

**Part III    Applications and Case Studies**

# Contributors

**Mehdi Behroozi**  Northeastern University, Boston, MA, USA

**Matthias Bender**  Research Center for Information Technology (FZI), Karlsruhe, Germany

**Burcin Bozkaya**  Sabanci University, Istanbul, Turkey

**Miguel Camacho-Collados**  Cabinet of the Secretary of State for Digital Advancement, Madrid, Spain

**John Gunnar Carlsson**  University of Southern California, Los Angeles, CA, USA

**Juan A. Díaz**  Universidad de las Américas Puebla, Cholula, Mexico

**Shakiba Enayati**  State University of New York, Plattsburgh, NY, USA

**Jörg Kalcsics**  The University of Edinburgh, Edinburgh, UK

**Hyun Kim**  University of Tennessee at Knoxville, Knoxville, TN, USA

**Kamyoung Kim**  Kyungpook National University, Daegu, South Korea

**Federico Liberatore**  Universidad Complutense de Madrid, Madrid, Spain

**Dolores E. Luna**  Universidad de las Américas Puebla, Cholula, Mexico

**Maria E. Mayorga**  North Carolina State University, Raleigh, NC, USA

**Juan G. Moya-García**  Linde México, Nuevo León, Mexico

**Osman Y. Özaltın**  North Carolina State University, Raleigh, NC, USA

**Federica Ricca**  Sapienza Università di Roma, Rome, Italy

**Roger Z. Ríos-Mercado**  Universidad Autónoma de Nuevo León, San Nicolás de los Garza, Mexico

**M. Angélica Salazar-Aguilar**  Universidad Autónoma de Nuevo León (UANL), San Nicolás de los Garza, NL, Mexico

**María G. Sandoval**  Universidad de las Américas Puebla, Cholula, Mexico

**Andrea Scozzari**  Università degli Studi Niccolò Cusano, Rome, Italy

**Begoña Vitoriano**  Universidad Complutense de Madrid, Madrid, Spain

**Seda Yanık**  Istanbul Technical University, Istanbul, Turkey

# Part I
# Introduction and Literature Reviews

# Chapter 1
# Research Trends in Optimization of Districting Systems

Roger Z. Ríos-Mercado

## 1.1 Motivation

The intent of this book is to present recent developments and insights on optimal territory design problems. In the literature, territory design can also be referred as districting or zone design. In particular emphasis is given to modeling aspects, theory, and algorithmic development of recent complex developments on districting and territory design by leading experts in the field. The book includes some literature surveys on particular areas of districting such as police patrolling, health care districting, and computational geometry methods, and successful case studies in political and sales force deployment districting.

The area of districting or territory design can be seen as a subfield of discrete optimization related to partitioning decisions. In a typical districting problem, a collection of basic or geographic units must be divided into districts or territories. This partition is not arbitrary but must meet a series of planning requirements depending on the specific application or context.

Although there is no such thing as "the Districting Problem" because each problem is different and has its own particular requirements that make it unique, there are certainly some criteria or requirements that are common to an important class of districting problems. For instance, criteria such as *compactness*, *unique assignment*, and *balance* are common to many districting problems. Other criteria such as *contiguity*, *similarity with existing plan* often appear as well. *Territory compactness* has to do with having territories formed by basic units that are close as possible from each other. This is typically achieved by minimizing a dispersion

R. Z. Ríos-Mercado (✉)
Department of Mechanical and Electrical Engineering, Graduate Program in Systems Engineering, Universidad Autónoma de Nuevo León, San Nicolás de los Garza, Mexico
e-mail: roger.rios@uanl.edu.mx

© Springer Nature Switzerland AG 2020
R. Z. Ríos-Mercado (ed.), *Optimal Districting and Territory Design*, International Series in Operations Research & Management Science 284, https://doi.org/10.1007/978-3-030-34312-5_1

function. In application such as political districting, dispersion functions that take into account the actual shape of the units are often used. *Unique assignment* means that each basic unit must be assigned to a single district. In other words, this feature assures a partition of the set of basic units. Exceptions to this rule can be found, for instance, in Fernández et al. [2] or Ríos-Mercado and Bard [10]. *Territory balance* implies that the total amount of "work" must be fairly distributed among districts. By "work" we mean whatever particular attribute or attributes are measured in each basic unit. Examples of this are population size, product demand, number of customers, workload, and so on. *Territory contiguity* or *connectivity* appears when there is an underlying graph representing adjacency between basic units, and assures that each territory must induce a connected subgraph. One example of this arises in political districting applications. *Similarity with existing plan* has to do with redistricting a current partition in such a way that is as similar as possible as the existing one. In commercial or distribution districting, for instance, keeping existing customer-driver relationships is often deemed as very important. Naturally, there is not a unique way of representing or modeling each of these aspects, thus a significant amount of research has been done studying different ways of dealing with these issues. Just to give an example, there are different ways of modeling territory connectivity. Some models are based on polynomial flow-based formulations [12], others are based on exponential amount of explicit connectivity constraints, handled by cut-generation algorithms [11]. Naturally, there are trade-offs that would depend on the particular districting application.

During the 1960–1980s, the main areas of work were dominated by mainly political districting and sales territory design. In the past 30 years or so, aside from these applications we have seen more studies on service-related districting, distribution/commercial territory design, and, more recently, districting in health care applications such as designing districts for location of emergency medical service (Chap. 3) or designing districts for efficient organ transplantation [3].

Now, in terms of solution methodologies, given the inherent computational complexity of most problems, it is not surprising that most of the research done for solving districting problems has been on heuristic and metaheuristic approaches. There is a clear practical impact that requires quick solutions implemented in practice. Nevertheless, some problems have special structure and properties that make them attractive for exact optimization schemes.

The reader can find a number of excellent surveys discussing many aspects of modeling, assumptions, solution methodologies, and applications of territory design and districting. For instance, Zoltners and Sinha [13] present a survey of models, solution approaches, and managerial insights for sales districting problems (1974–2004). They pay special attention to the economical impact that good territory alignment practices and processes have had over the years. Kalcsics et al. [6] present the first extensive literature review on models, methods, and applications for general districting problems. They discuss common features to a large class of districting problems and present a basic territory design model. They discuss in detail two approaches for this basic model: a classical location-allocation approach combined with optimal split resolution techniques and a new method based on computational geometry. They discuss extensions to the basic model and its integration into geographic information systems. Duque et al. [1] review almost four decades (1960s–2000s) of contributions on districting or supervised regionalization methods with main focus on spatially contiguous districts. The authors present a taxonomic scheme that classifies a wide range of regionalization methods into eight groups, based on the strategy applied for satisfying the spatial contiguity constraint. The paper includes a qualitative comparison of these groups in terms of a set of certain features, and a discussion of future lines of research for extending and improving these methods. Ricca et al. [9] present a complete literature survey on political districting highlighting modeling aspects from classical to current approaches. More recently, Kalcsics and Ríos-Mercado [5] present a wide overview and detailed discussion of typical criteria and requirements arising in territory design and how these have been modeled. The discussion includes an overview of many different application areas and the most relevant solution methodologies. Finally, in this book, Chaps. 2 and 3 present up-to-date literature reviews in two very hot areas such as police patrolling and health care districting, respectively, and Chap. 4 reviews computational geometry approaches for continuous-based districting.

## 1.2   Research Contributions

### 1.2.1   Part I: Introduction and Literature Reviews

The first part of this book contains a literature review and discussion on two very important areas of districting, namely police districting and healthcare districting. In Chap. 2, Liberatore, Camacho-Collados, and Vitoriano present a systematic literature review on police districting problems. They classify the main contributions in terms of model attributes and solution techniques employed. The chapter includes an annotated bibliography discussing the most relevant works on this area.

Chapter 3, by Yanık and Bozkaya, presents a review of literature of districting problems arising in health care. The health care districting problems are classified into three main areas: home care services, primary and secondary health care services, and emergency health care services. The chapter highlights modeling approaches, assumptions, and solution methods for each problem. The chapter ends with a discussion of several avenues of opportunity for future research areas.

Chapter 4, by Behroozi and Carlsson, presents a review of districting algorithms based on computational geometry. These algorithms contrast with other algorithms that are based on discrete network-based models. The chapter includes a discussion on how and when these type of algorithms can be applied and be more useful.

## 1.2.2 Part II: Theory, Models, and Algorithms

The second part of the book highlights recent advances on theory, modeling, and algorithms including mathematical programming and heuristic approaches. Chapter 5, by Díaz, Luna, and Sandoval, presents lower and upper bound procedures for a class of territory design problems that consider the minimization of a $p$-median problem dispersion function subject to planning requirements such as connectivity and balance with respect to one or more activity measures. The chapter also presents exact methods that use different linear programming relaxations.

Chapter 6, by Ricca and Scozzari, addresses political districting problems. The main focus of this chapter is on particular modeling aspects arising on several classes of political districting problems. Special care is given to the district connectivity requirement, which is essential in political districting applications. The chapter includes a discussion of the main contributions in the literature addressing this feature.

In Chap. 7, by Bender and Kalcsics, a multi-period service districting problem is addressed. This considers an important, but not commonly studied feature of many service districting applications consisting of customers requiring service under different frequencies. A consequence of this is that, in addition to the districting decisions, visiting schedules within the planning horizon must be decided as well. These decisions must take into account a fair workload balance for each service provider across all time periods and territory compactness. The chapter presents a MILP model, a discussion of its properties, and a development of a branch-and-price framework for solving the problem.

In Chap. 8, by Enayati, Özaltın, and Mayorga, an ambulance service districting subject to uncertainty is addressed. A two-stage stochastic mixed-integer programming model is presented. The proposed model suggests how to locate ambulances to the waiting sites in the service area, and how to assign a set of demand zones to each ambulance at different backup levels. The proposed stochastic service district design (SSDD) model enables quick response times by jointly addressing the location and dispatching policies in a stochastic and dynamic environment. The model maximizes the expected number of covered calls, while restricting the workload of each ambulance. An interesting feature is that the proposed model can be optimized offline. The chapter includes an empirical assessment of the model through a discrete-event simulation and comparison with two baseline policies. The results indicate significant improvements in many related metrics.

## 1.2.3 Part III: Applications and Case Studies

The third part of the book contains successful applications in real-world districting cases. Chapter 9, by Kim and Kim, highlights the need of determining the location of polling facilities and polling stations tailored to the regulations of the voting process of South Korea by means of a spatial optimization approach. In the spatial model, a utility cost function that captures the effects of distance and preference, such as that based on pre-knowledge of or experience with existing facilities, is formulated. The chapter includes a case study with real-world data in Seoul, Korea. The numerical results indicate the need to relocate the existing polling facilities, merge certain precincts, and adjust existing boundaries of precincts to enhance the efficiency of administration of the voting process.

Chapter 10, by Moya-García and Salazar-Aguilar, focuses on a sales territory design application. In this particular problem, a sales force team is in charge of performing recurring visits to customers, where each territory is assigned to a sales representative with the aim of establishing long-term personal relationship with the customers. At the strategic level, the decision maker must partition the set of customers into sales territories and at the tactical level, the daily routes (schedule of visits) of the sales representatives must be planned. Balanced territories allow better customer coverage and balanced workload. Additionally, efficient routes allow to perform more visits and to reduce travel time. The chapter presents a heuristic approach for this problem. The method is applied and assessed in two case studies arising in a Mexican firm. Computational findings indicate the effectiveness of the proposed heuristic, producing high-quality solutions.

## 1.3 Closing Remarks

Although the models and techniques presented in this book are applied to specific situations, it is evident that many of these techniques and ideas can be applied or adapted to other more complex problems.

Just as an example, there are a few works that consider districting under uncertainty [4, 7, 8]. Many of the scenario-based and/or decomposition approaches rely on knowing how to efficiently solve or handle deterministic subproblems. The ideas exposed in this volume can certainly be of very high value when devising such approaches.

There are also applications where districting decisions (which are essentially tactical decisions) must be taken along with operational decisions (such as scheduling, routing, and so forth). Chapter 7 is a good example of this. Again, efficient decomposition approaches that rely on effective districting solution algorithms can be studied under these ideas.

## References

1. Duque, J.C., Ramos, R., Suriñach, J.: Supervised regionalization methods: a survey. Int. Reg. Sci. Rev. **30**(3), 195–220 (2007)
2. Fernández, E., Kalcsics, J., Nickel, S., Ríos-Mercado, R.Z.: A novel maximum dispersion territory design model arising in the implementation of the WEEE-directive. J. Oper. Res. Soc. **61**(3), 503–514 (2010)

3. Gentry, S., Chow, E., Massie, A., Segev, D.: Gerrymandering for justice: redistricting U.S. liver allocation. Interfaces **45**(5), 462–480 (2015)
4. Haugland, D., Ho, S.C., Laporte, G.: Designing delivery district for the vehicle routing problem with stochastic demands. Eur. J. Oper. Res. **180**(3), 997–1010 (2007)
5. Kalcsics, J., Ríos-Mercado, R.Z.: Districting problems. In: Laporte, G., Nickel, S., Saldanha da Gama, F. (eds.) Location Science, 2nd edn., chap. 25. Springer, Berlin (2020)
6. Kalcsics, J., Nickel, S., Schröder, M.: Towards a unified territorial design approach: applications, algorithms, and GIS integration. TOP **13**(1), 1–56 (2005)
7. Lei, H., Laporte, G., Guo, B.: Districting for routing with stochastic customers. EURO J. Transport. Logist. **1**(1–2), 67–85 (2012)
8. Lei, H., Wang, R., Laporte, G.: Solving a multi-objective dynamic stochastic districting and routing problem with a co-evolutionary algorithm. Comput. Oper. Res. **67**, 12–24 (2016)
9. Ricca, F., Scozzari, A., Simeone, B.: Political districting: from classical models to recent approaches. Ann. Oper. Res. **204**(1), 271–299 (2013)
10. Ríos-Mercado, R.Z., Bard, J.F.: An exact algorithm for designing optimal districts in the collection of waste electric and electronic equipment through an improved reformulation. Eur. J. Oper. Res. **276**(1), 259–271 (2019)
11. Salazar-Aguilar, M.A., Ríos-Mercado, R.Z., Cabrera-Ríos, M.: New models for commercial territory design. Netw. Spat. Econ. **11**(3), 487–507 (2011)
12. Shirabe, T.: Districting modeling with exact contiguity constraints. Environ. Plann. B. Plann. Des. **36**, 1053–1066 (2009)
13. Zoltners, A.A., Sinha, P.: Sales territory design: thirty years of modeling and implementation. Market. Sci. **24**(3), 313–331 (2005)

# Chapter 2
# Police Districting Problem: Literature Review and Annotated Bibliography

Federico Liberatore, Miguel Camacho-Collados, and Begoña Vitoriano

**Abstract** The police districting problem concerns the efficient and effective design of patrol sectors in terms of performance attributes. Effectiveness is particularly important as it directly influences the ability of police agencies to stop and prevent crime. However, in this problem, a homogeneous distribution of workload is also desirable to guarantee fairness to the police agents and an increase in their satisfaction. This chapter provides a systematic review of the literature related to the police districting problem, whose history dates back to almost 50 years ago. Contributions are categorized in terms of attributes and solution methodology adopted. Also, an annotated bibliography that presents the most relevant elements of each research is given.

## 2.1 Introduction

We are the lucky witnesses of a revolution taking place in the way police agencies work. In the last decade, we have seen the rise of predictive policing, i.e., the use of mathematical and statistical methods in law enforcement to predict future criminal activity based on past data. Its importance has been even recognized by Time magazine that in November 2011 named predictive policing as one of the 50 best inventions of 2011 [15].

Apart from crime forecasting, mathematics still have a major role to play in policing and its various disciplines can help by giving police agency a new edge

F. Liberatore (✉) · B. Vitoriano
Department of Statistics and Operational Research and Institute of Interdisciplinary Mathematics, Universidad Complutense de Madrid, Madrid, Spain
e-mail: fliberat@ucm.es; bvitoriano@mat.ucm.es

M. Camacho-Collados
Artificial Intelligence Area Organization, Cabinet of the Secretary of State for Digital Advancement, Madrid, Spain
e-mail: mcamachoc@mineco.es

© Springer Nature Switzerland AG 2020
R. Z. Ríos-Mercado (ed.), *Optimal Districting and Territory Design*, International Series in Operations Research & Management Science 284,
https://doi.org/10.1007/978-3-030-34312-5_2

in the fight against crime. This is also recognized by the RAND Corporation and the National Institute of Justice of the United States (NIJ). In fact, both these prestigious institutions acknowledge the need for taking a step forward and developing explicit methodologies and tools to take advantage of the information provided by predictive policing models to support decision makers in law enforcement agencies [24].

Districting models are a natural way to make use of crime forecasts to design police districts tailored to the criminal behavior of an area. During most of the twentieth century, police districts were drawn by police officers on a road map with a marker, just by following the major streets in the area, or according to neighborhood perimeters, without considering workload efficiency or balance [3]. The first model for the design of police district was formulated by Mitchell [23], almost 50 years ago. Since then, a number of mathematical optimization models have been proposed and the police districting problem (PDP) was born. The objective of the PDP is to partition the territory under the jurisdiction of a Police Department in the "best possible way." PDP models normally consider several attributes, such as time, cost, performance, and other topological characteristics.

Geographic information systems (GIS), thanks to their ability to represent and manipulate geographical data using a reasonable amount of computational time, gained popularity among both academics and practitioners which started to contemplate the possibility of adopting automatic methodologies for the definition of police districts [28]. However, studies integrating GIS and sophisticated mathematical modeling for police districting remain a rarity [3], and the design of police districts is often based on the experience and intuition of few experts. Nevertheless, the importance of a balanced definition of the police districts is unquestioned and the implementation of decision-aid tools for the allocation of police resources has proven to be extremely beneficial, as shown by the numerous papers presented in the academic literature in the last decades [13]. In fact, all the researches report a dramatic improvement in workload distribution compared to hand-made districts which, in turn, results in enhanced performance and efficiency.

In this chapter we provide a general definition for the PDP and analyze in detail the literature related to this topic. The PDP is formally presented in Sect. 2.2. In Sect. 2.3, previous contributions in this line of research are categorized in terms of attributes considered and methodologies and approaches adopted for the problem solution. Next, an annotated bibliography is provided, where a brief description of the most salient points of each research is given (see Sect. 2.4). Conclusions are given in Sect. 2.5.

## 2.2 The Police Districting Problem

In the USA, the territory under the jurisdiction of a police department is partitioned according to a hierarchical structure constituted by command districts (or precincts), patrol sectors (or beats), and reporting districts (or r-districts or blocks). Command districts host the headquarters where the commanding officer supervises the operations and are fractioned into patrol sectors. Patrol sectors have one or more cars assigned which patrol the area and attend to the calls for service that originate in the sector. R-districts represent the smallest geographical unit for which statistics are kept and are, *de facto*, the atomic element in the hierarchy. Sarac et al. [26] explain that r-districts can coincide with census block groups as it is convenient to do so for administrative reasons. The territorial structure in Europe is not as homogeneous as in the USA, as it depends on the country or the region considered. However, a hierarchal structure similar to the one adopted in USA is predominant.

The PDP concerns the optimal grouping of blocks into "homogeneous" patrol sectors in such a way that all the territory is partitioned and that no sector is empty. It is desirable for the patrol sectors to be connected and topologically efficient (e.g., compact). In fact, the car(s) assigned to the patrol sector should respond to all the incidents taking place in the area and, therefore, topologically efficient sectors would result in a diminished travel time and, in turns, in a higher operational effectiveness. Normally, if all the cars in a sector are busy responding calls, a car from a neighboring sector has to attend the incoming calls. This generally leads to a domino effect where cars are pulled from their area to another, leaving the patrol sector unattended and, therefore, more susceptible to criminal incidents (as pointed out by Mayer [21]). In the light of this scenario, balanced workload among the districts and enforcement of a maximal response time become of primary importance.

Figure 2.1 shows a crime heat-map for a Saturday night shift in the Central District of Madrid, Spain, and the corresponding PDP solution. The borders of the census districts have been plotted in black, the streets in gray and each patrol sector is represented by a different color. It can be observed that the sector design is adjusted to provide an equitable territory partition among the beats.

A generic formulation for the PDP is given in the following.

$$\min \quad obj\,(P) \tag{2.1a}$$

$$\text{subject to} \quad \emptyset \notin P \tag{2.1b}$$

$$\bigcup_{A \in P} A = N \tag{2.1c}$$

$$A \cap B = \emptyset \qquad A, B \in P \,|\, A \neq B \tag{2.1d}$$

$$|P| = p \tag{2.1e}$$

$$Conn\,(A) = 1 \qquad A \in P \tag{2.1f}$$

The model optimizes a certain objective function evaluated on a partition $P$. Constraints (2.1b)–(2.1d) represent the conditions held by a partition $P$ defined on the set of blocks $N$: $P$ should not contain the empty set $\emptyset$ (2.1b), the partition covers entirely $N$ (2.1c), and the sectors are pairwise disjoint (2.1d). The restriction (2.1e) concerns the partition cardinality and enforces the number of patrol sectors to be exactly $p$. Finally, condition (2.1f) regards the geometry of the patrol sectors.

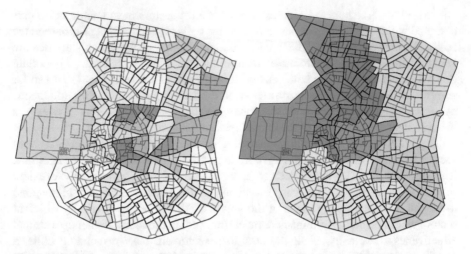

**Fig. 2.1** Crime heat-map for a Saturday night shift in the Central District of Madrid, Spain (left), and sample solution obtained by a PDP model. The borders of the census districts have been plotted in black, the streets in gray and each patrol sector is represented by a different color. Source: Liberatore and Camacho-Collados [20]

*Conn* (*A*) is an indicator function that equals 1 when *A* is connected and zero otherwise. Therefore, this constraint establishes that only connected partition blocks are feasible. This condition implies that an agent cannot be assigned to a patrol sectors spanning two or more separate areas of the city.

In its most basic form, the PDP is not different from any districting problem. Of course, the basic formulation can be enriched with specific constraints and conditions. In particular, the different PDP presented in the literature includes attributes and parameters that represent the idiosyncrasies of the policing context, as explained in the following section.

## 2.3   Literature Review

This section is devoted to an analysis and categorization of the attributes and methodologies adopted in the literature on the PDP. A summary of the findings is illustrated in Table 2.1, where the most salient characteristics of each contribution are presented.

### 2.3.1   Attributes

While analyzing the existing literature on the PDP, certain basic features common to all contributions emerged. In fact, all the applications considered include measures

**Table 2.1** Mapping of attributes considered and methodology adopted, by article

| Reference | Attributes | | | | Methodology |
|---|---|---|---|---|---|
| | Workload | Response time | Geometry | Other | |
| Mitchell [23] | Expected service time, expected travel time | Expected travel time | | | Modified clustering heuristic |
| Bodily [2] | Fraction of time spent servicing calls | Average travel time | | | Utility theory |
| Benveniste [1] | Probability of a server being found busy | Total expected station-alarm distance | | | Stochastic optimization |
| Sarac et al. [26] | Homogeneity in terms of population and call volume | | Area, compactness, contiguity | Easy access to demographic data, suitable for use by other agencies, and respect of existing district boundaries | Redefinition according to census blocks |
| D'Amico et al.[13] | Utilization of servers | Queuing response time and travel time | Size, compactness, contiguity, convexity | | Queuing model and simulated annealing |
| Curtin et al. [11] | Maximum incident load per sector | Maximum service distance | | | GIS and mathematical programming optimization |
| Kistler [18] | Total hours worked, number of calls, population | Average travel time | Area, total length of streets, compactness | Boundaries of gang territories, city council wards, neighborhood associations, and Air Force Base | GIS |
| Curtin et al. [12] | Maximum incident load per sector | Maximum service distance | | Backup coverage | GIS and mathematical programming optimization |

(continued)

**Table 2.1** (continued)

| Reference | Attributes | | | | Methodology |
|---|---|---|---|---|---|
| | Workload | Response time | Geometry | Other | |
| Zhang and Brown [29] | Homogeneity in terms of average travel time and response time | Average travel time | | | GIS and agent-based simulation |
| Zhang and Brown [30] | Homogeneity in terms of average travel time and response time | Average travel time | | | Simulated annealing and discrete-event simulation |
| Zhang and Brown [31] | Homogeneity in terms of average travel time and response time | Average travel time | | | GIS, experimental design methods, agent-based simulation, and discrete-event simulation |
| Bucarey et al. [4] | Homogeneity in terms of sector demand | | Sector boundaries | Prevention demand component | Mathematical programming optimization |
| Camacho-Collados et al. [6] | Homogeneity in terms of area, risk, isolation, and diameter | | Diameter. Sectors must be connected and convex. | Isolation, i.e., how far the sector is from other sectors | Mathematical programming optimization |
| Cheung et al. [9] | | | | | Mathematical programming optimization |
| Chow et al. [10] | | | | | Mathematical programming optimization |
| Liberatore and Camacho-Collados [20] | Homogeneity in terms of area, risk, isolation, and diameter | | Diameter. Sectors must be connected and convex. | Isolation, i.e., how far the sector is from other sectors | Mathematical programming optimization |
| Piyadasun et al. [25] | Crime-severity weighted distance. Homogeneity measured using Gini index | | Rectangular area needed to cover the whole sector. Isoperimetric quotient | | Clustering heuristic |

for workload, response time, and geometrical properties of the districts. Neverthe-less, the implementation shows significant variations. Differently from Kalcsics and Schröeder [17], the denomination "attributes" has been adopted instead of "criteria" with the aim to provide a more generic framework that classifies all the relevant characteristics of a PDP solution, regardless of whether they are optimized in the objective function, or expressed as constraints.

### 2.3.1.1  Workload

Given the complex nature of police procedures and operations, and the great variability of tasks that an agent can undertake, defining workload could be complicated. Bruce [3] poses a number of questions that can help to clarify what to consider as part of the workload. Albeit difficult, an accurate definition of workload is desirable as it ensures homogeneity in terms of quality and speed of service, and equalizes the burden on police officers [2].

In the literature on the PDP, different definitions of workload have been adopted. Mitchell [23] computes the workload as the sum of the total expected service time and the total expected travel time. Curtin et al. [11, 12] use the number (or frequency) of calls (or incidents) occurring at each district as a proxy for the workload. As different calls can have different service time, some authors reckon that this measure is too naïve as it might produce unbalanced patrol districts. For Bodily [2] and D'Amico et al. [13] workload is defined as the fraction of working time that an agent spends attending to calls. An equivalent measure is proposed by Benveniste [1]. Given the stochastic nature of her model, workload is measured in terms of probability of a patrol car being busy. Once the probability is known, the time spent attending and answering calls can be easily calculated. More recently, workload has been defined as a combination of different characteristics. Sarac et al. [26] aggregate population and call volume. Kistler [18] makes use of the convex combination of total hours worked (i.e., from dispatch to call clearance), number of calls, and population. Zhang and Brown [29–31] consider both average travel time and response time. Camacho-Collados et al. [6] and Liberatore and Camacho-Collados [20] define workload as the weighted combination of multiple attributes: area, risk, isolation, and diameter. Finally, Piyadasun et al. [25] define the workload as the sum of the distance of the district center to its crime points, weighted by the severity of the crime. Interestingly, equality in the distribution of the workload among patrol sectors is measured using the Gini coefficient.

### 2.3.1.2  Response Time

Response time is an important performance measure representing the time between the arrival of a call for service and the arrival of a unit at the incident location. According to Bodily [2], the reduction of the response time results in a number of beneficial effects such as:

- Increased likelihood of intercepting a crime in progress.
- Deterrent effect on criminals.
- Increased confidence of citizens in the police.

Generally speaking, most of the authors consider exclusively travel times [2, 18, 23, 29–31] or travel distances [1, 11, 12]. The only study considering queuing effect is by D'Amico [13], where the authors apply an external model—PCAM [7, 8]—to compute the total response time including calls queuing time and travel time to the incident location.

### 2.3.1.3 Geometry

In 1812, Albright Gerry, the Governor of the State of Massachusetts at the time, manipulated the division of his state and proposed a salamander-shaped district to gain electoral advantage, leading to the expression "gerrymandering" (resulting from merging "Gerry" and "salamander"). Since then, designing electoral districts having certain geometric properties is of primary importance to ensure neutrality and prevent political interference in the districting process.

In the context of the PDP, geometric attributes are still relevant for efficiency (e.g., establishing boundaries that would be easy to patrol and would not frustrate patrol officers) and for administrative reasons (e.g., coordination with other agencies). A number of researches explicitly include geometric properties in the design process, as the properties of compactness [6, 13, 18, 20, 26], contiguity [6, 13, 20, 26], and convexity [6, 13, 20] are generally obtained as a consequence of optimizing the travel distance or the travel time. Also, the district area is considered in all the mentioned works. Camacho-Collados et al. [6] and Liberatore and Camacho-Collados [20] achieve compactness by minimization of the sectors' diameter. Additionally, Kistler [18], Camacho-Collados et al. [6], and Liberatore and Camacho-Collados [20] include the total length of the streets in a patrol sector.

### 2.3.1.4 Other Attributes

Recently, a number of attributes that do not fall into any of the previous categories have been introduced by some works. These attributes generally try to capture complex real-world requirements. The Buffalo Police Department needed to redesign r-districts in such a way that the existing district boundaries would be respected, and the access to demographic data as well as the use by other agencies would be easy [26]. The Tucson Police Department needed to consider the boundaries of gang territories, city council wards, neighborhood associations, and the Davis–Monthan Air Force Base [18]. Curtin et al. [12] introduce backup coverage (i.e., multiple coverage) of incident locations. Camacho-Collados et al. [6] and Liberatore and Camacho-Collados [20] define an isolation attribute that expresses how far the sector is with respect to the others. The rationale is that a more isolated sector can receive less support than a more central one. Finally, Bucarey et al. [4] propose a prevention demand component that represents the need for police resources used for preventive patrols. This component is calculated as the maximum between three factors, each multiplied by a scaling coefficient. The factors considered are reported crime, population, and total kilometers of roads in the sector.

## 2.3.2   Methodologies and Approaches

Many districting approaches have appeared in the literature. In this subsection, the contributions are categorized according to the methodology adopted and their main characteristics are presented.

### 2.3.2.1   Optimization Models

According to Kalcsics and Schröeder [17], the first mathematical program for the Districting Problem was proposed by Hess et al. [16], and regarded the neutral definition of political district. Since then, a large number of models have been presented, mostly in the context of location analysis. Mitchell [23] defines a set partitioning model that considers minimizing the expected distance inside of each subset and equalizing workload among all the subsets. Curtin et al. [11, 12], propose maximal covering models. Cheung et al. [9] and Chow et al. [10] consider both the $p$-median problem and the maximum coverage problem. Bucarey et al. [4] formulate their problem as an enriched $p$-median model. Finally, Camacho-Collados et al. [6] and Liberatore and Camacho-Collados [20] introduce a multi-criteria police districting problem that provides a balance between efficiency and workload homogeneity, according to the preferences of a decision maker. It is important to notice that all these formulations, due to the size of the application context, are often solved heuristically. This is also the case for Piyadasun et al. [25] that, despite not presenting any mathematical formulation for their problem, solve it by means of an *ad-hoc* clustering heuristic.

A different perspective is adopted by Benveniste [1] and D'Amico et al. [13], where patrol cars and agents are modeled as servers in a stochastic model. In particular, Benveniste [1] proposes a stochastic optimization model, while D'Amico et al. [13] include a queuing model inside of a simulated annealing algorithm to compute response times that incorporate queuing effects.

### 2.3.2.2   Geographic Information Systems (GIS)

The first application of geography to crime analysis dates back to 1829, when the Italian geographer Adriano Balbi and the French lawyer André-Michel Guerry drew three maps of crimes in France, highlighting geographic patterns of crime and their relations [19]. Since then, the marriage between geography and criminology gave birth to numerous methodologies. When the GIS were developed, their implementation in law enforcement agencies and crime research was only natural, and in the last decade we are assisting to an extremely rapid growth of adoption, supported by the promotion of the NIJ (United States National Institute of Justice). For a review of GIS application to crime research the reader is referred to the overview by Wang [28].

According to this trend, the last works on the PDP are developed in the framework of the GIS. Kistler [18] uses a GIS to redesign the Tucson Police Department districts. Most commercial GIS can be extended to integrate optimization routines. Curtin et al. [11, 12] use GIS in conjunction with a maximal covering model. Finally, Zhang and Brown [29–31] implement agent-based simulation, and discrete-event simulation inside of a GIS.

### 2.3.2.3   Other Methods

Two studies adopted approaches that do not fall into any of the other categories. Bodily [2] devises a decision model based on utility theory to achieve the best solution with respect to the surrogate utility of three interest groups. The work by Sarac et al. [26] is an example of the idiomatic expression "simpler is better." After attempting to redesign r-districts by using a multi-criteria set partitioning model, the authors realized that census blocks satisfied all the requirements. It is important to notice that their approach is successful because of the specific requisites the Buffalo PD had on the r-district configuration (e.g., easy access to demographic data, suitable for use by other agencies).

## 2.4   Annotated Bibliography

In the following, an annotated bibliography providing a description of the most salient points, achievements, and characteristics of the most relevant contributions in the PDP is given. The contributions are presented in chronological order.

### 2.4.1   Mitchell [23]

In his seminal work, Mitchell presents a mathematical formulation for the problem of designing police patrol sectors. The model is based on the assumption that incident distribution is known and that each call is serviced by the nearest available units. Multiple incident types are considered. Each type is characterized by a service time and a vector of weights that define the importance of the incident being attended by a specified number of units. The model minimizes the total expected weighted distance. Also, the workload, defined as the sum of the expected service time and the expected travel time, is equalized across the sectors. The problem is solved by means of an adapted clustering heuristic and applied to incident data for Anaheim, California. The solution improves the sector plan adopted at the time.

## 2.4.2   Bodily [2]

Bodily proposes a decision model based on utility theory, which makes use of the preferences of three interest groups in the design process of police sectors: citizens (minimize travel time, equalize travel time), administrators (minimize travel time, equalize travel time, and equalize workload), and service personnel (equalize workload). The problem is solved by a local search algorithm that transfers one block from one sector to another, so that the greatest improvement in terms of surrogate utility is achieved. The algorithm stops when no improvement is possible.

## 2.4.3   Benveniste [1]

The author presents a stochastic optimization problem for the combined zoning and location problem for several emergency units. Namely, the problem involves the division of an area in sectors, the definition of location for the servers, and a set of rules, assigning for service an alarm to a list of servers in order of preference. The objective function considered minimizes the total expected distance between the alarm and the first available server. Stochastic alarms rates, alarms spatial density functions, and probabilities that the servers are busy are considered. The resulting model is a non-linear program. The author proposes an approximation algorithm. An equal workload case is also examined and solved.

## 2.4.4   Sarac et al. [26]

The authors describe a study undertaken to reconfigure the police reporting districts used by the Buffalo Police Department. The following districting criteria were considered:

- Homogeneity in terms of population, area, and call volume.
- Geometrical properties such as compactness and contiguity.
- Feasibility with respect to existing boundaries of five police districts.
- Easy access to demographic data for each district.
- Suitability for use by other agencies.

Initially, the authors formulated the problem as a multi-objective set partition problem which proved incapable to solve the real-world size problem at hand. Subsequently, a practical approach has been proposed: basically, the new districts were defined according to the census block groups that intrinsically present most of the desired characteristics (homogeneity in terms of population, compactness, contiguity, easy access to demographic data, and suitable for use by other agencies). With minor modifications, this solution proved to be very effective.

### 2.4.5   D'Amico et al. [13]

The authors solve a PDP subject to constraints of contiguity, compactness, convexity, and equal size. The novelty of the model lies in the incorporation of queuing measures to compute patrol offices workloads and response times to calls for service, computed by external software, PCAM. PCAM optimizes a queuing model for deployment of police resources, providing the optimal number of cars per district. The authors solve the problem by means of a simulated annealing algorithm that iteratively calls the PCAM routine. At each step, the neighborhood is determined by a simple exchange procedure that takes into account the following constraints:

- The average response time per district is bounded from above.
- The ratio of the size of the largest and smallest districts is bounded from above.
- Sectors must be connected.
- The ratio of the longest Euclidean path and the square root of the area in each sector is bounded from above to preserve compactness.
- Sectors must be convex.

The algorithm is applied to a real-world case for the Buffalo Police Department. The following objectives were considered: minimization of the maximum workload (by decremental bounding constraining) and minimization of the maximum average response time.

### 2.4.6   Curtin et al. [11]

The authors apply a covering model to determine police patrol sectors. The covering model defines the centers of police patrol sectors in such a way that the maximum number of (weighted) incidents is covered. An incident is considered to be covered if it lies within an acceptable service distance from the center of a patrol sector. The model is integrated in a GIS and applied to a case study on the City of Dallas, Texas.

In the final part of this chapter, the authors present a number of possible refinements to their model, including a maximum workload restriction (in terms of number of weighted incidents).

## 2.4.7   Scalisi et al. [27]

The issue of Geography & Public Safety presents numerous articles by police analysts describing their experiences with police redistricting within their police department.

- Bruce [3]: C. Bruce, President of the International Association of Crime Analysts, poses some interesting questions that an analysts should answer to determine how workload should be measured.
- Kistler [18]: A. Kistler, from the Tucson Police Department, devises a district evaluation measure built as the weighted sum of the following criteria: total hours worked, number of call responses, average response time, total length of all streets within the division, area of the division, and population. TPD staff used a GIS in combination with a software program called Geobalance to manually design alternative districting configurations. Future evaluations of the implemented plan showed that the projected workload ratios were reliable and realistic.
- Douglass [14]: J. Douglass, from the Overland Park Police Department, explains how the introduction of a new real-time deployment paradigm, based on criminal hot-spots identification and treatment, had been implemented in the department. Unfortunately, no long-term statistical analysis was available at the time the article was written.
- Mayer [21]: A. Mayer, from the East Orange Police Department, reports a similar strategy. In fact, the East Orange Police Department implemented a geographical technology called tactical automatic vehicle locator (TAC-AVL). TAC-AVL allows for GPS tracking, visualization on a map, and recording of information regarding patrol cars and incidents. This tool has been coupled with a new deployment strategy that allows for the introduction of impact, resource, response, and conditions cars to backup understaffed zones of the jurisdiction.
- Mielke [22]: P. Mielke, from Redlands Police Department, explains how to use ESRI districting tool, a free extensions for ESRI ArcGIS that allows creating new police districts in a city or region.
- Other successful applications of geographical technologies to police redistricting have been reported from Austin PD and Charlotte-Mecklenburg PD.

## 2.4.8    Curtin et al. [12]

Following Curtin et al. [11], the authors extend the covering model to include backup coverage (e.g., multiple coverage of high priority locations). The resulting model is bi-objective in nature. The authors propose a single objective model that maximizes the priority weighted coverage (each time a location is covered is accounted for separately), while ensuring a minimum covering level in terms of priority weighted number of locations covered (each covered location is accounted for only once). The model is tested with the police geography of Dallas, Texas, and refinements on the model are proposed (e.g., maximum workload per patrol sector).

## 2.4.9    Wang [28]

The author takes us on a journey across the main application areas of GIS in police practices. Among the various applications, Wang mentions the possibility of using GIS as a police force planning tool. Namely, he refers to hot-spot policing and police districting. Concerning the latter, Wang identifies three main objectives: meeting a response time threshold, minimizing the cost of operation, and balancing workload across districts. The author mentions the work by Curtin et al. [11, 12] and states that future works in this area should explore other goals, such as minimizing total cost (response time), minimizing the number of sectors (dispatch centers), maximizing equal accessibility, or a combination of multiple goals.

## 2.4.10    Zhang and Brown [29]

In this work a parameterized redistricting procedure for police patrols sectors is proposed. The methodology consists of a heuristic algorithm that generates alternative districting configurations. Next, the configurations are evaluated in terms of average response time and workload. With this aim, an agent-based simulation model was implemented in a GIS. The location and times of the incidents taking place at each sector are modeled by an empirical distribution based on real incident data. Finally, the procedure identifies the set of non-dominated solutions. The methodology has been tested on a case study based on the Charlottesville Police Department, VA, USA.

## 2.4.11   Zhang et al. [32]

The focus of this research is the evaluation of three different methods for scoring police districting designs: a closed form probability based approach, a discrete-event simulation based on hypercube models for spatial queuing systems, and an agent-based simulation model. The scoring measures are evaluated on designs generated using the methodology presented in Zhang and Brown [29]. According to the authors, the three methods provide similar evaluations of the districting plans when the emergency response system is not stressed. However, in the face of high system stress, only the agent-based simulation model is capable of accurately representing the significant changes in the workload scores due to complex behaviors of the system such as cross-boundary support, that is, when an agent assigned to a district services a call for service in another district.

## 2.4.12   Zhang and Brown [30]

The research presented in this paper focuses on the definition of a simulated annealing algorithm for the problem of finding optimal police patrol districting designs. The optimization procedure makes use of a discrete-event simulation to evaluate the solutions according to multiple criteria, such as average response time and workload variation among sectors. Districting designs are generated using a simulated annealing procedure. In this procedure two different neighborhoods are compared. In the first one, only changes of one block are allowed. The second one consists of a cutting and merging process that allows for more significant changes. The authors show empirically that the second approach uses fewer iterations to reach good solutions and is, therefore, preferable when used in conjunction with discrete-event simulation.

## 2.4.13   Zhang and Brown [31]

In this paper, Zhang and Brown extend their previous research in Zhang and Brown [29]. The main changes with respect to the previous contribution are the following. First, both discrete-event simulation and agent-based simulation are considered. The former is more computational efficient while the latter is more precise. Second, an iterative searching procedure is used to optimize the parameters of the districting algorithm, instead of adopting a completely randomized approach. The authors propose using experimental design methods to explore the parameter space, but classical metaheuristics, such as simulated annealing and genetic algorithms, could be used as well. The methodology is tested on real data provided by the Charlottesville Police Department, VA, USA.

## 2.4.14 Bucarey et al. [4]

In this paper the authors define a variant of the classical $p$-median problem to tackle the problem of defining balanced police sectors. The model is designed keeping in mind the requirements of the Chilean National Police Force, but can be applied to any country. The model proposed enriches the classical $p$-median in a number of ways. First, it enforces bounds on the demand of each sector. The bounds can be specified according to the acceptable percentage of deviation from the average demand in order to ensure homogeneity. The objective function is defined as the weighted sum of three terms. The first one is the sum of the blocks distances to the associated median. The second term enforces compactness by considering the sum of a measure of the sectors' boundaries size. The function measuring the boundaries is non-linear in nature and is approximated by a piece-wise linear function. Finally, the third term represents the prevention demand component associated with a block. This component is defined as the maximum of three factors: the length of roads in the block, the amount of reported crime in the block, and the population of the block. The model is solved on a realistic case study considering 1266 blocks and up to $p = 7$ neighbors. Due to the size of the problem, the model is solved by means of a location-allocation heuristic algorithm.

## 2.4.15 Camacho-Collados et al. [6]

This paper presents the multi-criteria police districting problem (MC-PDP), a multi-criteria optimization model for partitioning the territory under the jurisdiction of a Police Department into sectors. The goal is the automatic definition of sectors that adapts to a particular shift. Initially, the territory is divided into a square grid. Each cell of the grid (which represents a block) is characterized by a crime risk, representing the expected crime, and an area, representing the total street length. A feasible design requires the sectors to be connected and convex. The workload for each sector is computed as the weighted sum of different factors: area, risk, isolation (i.e., how far the sector is from other sectors), and diameter. The objective function minimizes the weighted combination of the total workload and of the maximum workload. Assigning more weight to the first term results in solutions that are more efficient, while favoring the second term provides solutions that are more equal in terms of workload distribution. The model is solved by means of a multi-start randomized local search algorithm. The algorithm is tested on a real dataset including data from Central District of Madrid, Spain. A comparison with the configuration currently adopted by the Spanish National Police shows how this is suboptimal compared to the solutions found by the algorithm.

## 2.4.16   Camacho-Collados and Liberatore [5]

In this article, the authors embed the model and algorithm presented in Camacho-Collados et al. [6] within a decision support system for predictive police patrolling. The decision support system combines predictive policing features with the MC-PDP for the automatic definition of police districts that are tailored to the requirements of future shifts. The authors tackle the problem of adequately describing crime events in which the time and the location of the incidents are indeterminate. Previous research on the subject only contemplated temporally indetermined crimes.

## 2.4.17   Cheung et al. [9]

In this paper, a police force deployment optimization framework is proposed. The framework is comprised of two optimization problems solved in sequence: a police location problem and a patrolling area problem, respectively. The first problem is tackled as a $p$-median problem where nodes represent centroids generating crime and/or feasible locations for police facilities and the distances are weighted by the crime rate associated to the node. In the second problem, given the locations of police stations, the model determines the patrol area of the police force such that the total covered area of the police force is maximized. This is obtained by means of a maximum coverage problem. The framework is applied to a case study on the Greater London Area.

## 2.4.18   Chow et al. [10]

The authors apply classical operational research models—i.e., the $p$-median and the maximum coverage problems—for the location of $p$ police facilities. Travel costs (distances) and crime generated at each location are considered. The solutions to these problems provide a definition of police districts, that is crisp for the $p$-median problem (i.e., a location is always assigned to the closest facility) and fuzzy for the maximum coverage problems (i.e., a location is assigned to all the facilities that "cover" it, that is, the distance between the location and the facility is inferior to a predefined distance). The algorithms are tested on data representing the crime rate in January 2014 for all wards (i.e., blocks) in Greater London.

### 2.4.19    Liberatore and Camacho-Collados [20]

The focus of this article is the extension of the MC-PDP [6] to general graph structures. This allows for increased versatility in terms of applicability of the model. With respect to the model, the same criteria (i.e., area, risk, isolation, and diameter) and objective function are considered. However, the authors defined an efficient and practical condition for set convexity derived from the classical definition of convexity in graphs. In terms of solution methodologies, the authors propose three local search algorithms for the MC-PDP on a graph: simple hill climbing, steepest descent hill climbing, and Tabu search. Thanks to its ability to escape from local optima, the Tabu search algorithms find solutions that are on average better than the other two methodologies.

### 2.4.20    Piyadasun et al. [25]

Piyadasun et al. [25] propose a multi-step heuristic procedure that clusters census blocks into sectors. Initially, crimes are assigned to census blocks and the crime-weighted centroid of each census block is identified. Distances are determined using actual road distance. Then, $k$ non-contiguous centroids are chosen as the sector seeds. Next, the sectors are grown by adding a census block to a single sector at each iteration. The census block to be added is chosen in such a way that the resulting sector is as compact and efficient as possible. This is achieved by considering both the distance between the census block and the sector center (close census blocks are preferred) and the increase in terms of minimum rectangular area needed to cover the whole sector after the census block is added to it (small increases are preferred). The algorithm has been applied to crime data for the San Francisco County, CA, USA, corresponding to years from 2003 to 2015. The performance of the solutions obtained has been evaluated considering workload distribution, compactness of the districts, and patrol car response time. The workload definition used by the authors considers the number of calls for service in a district, as well as their severity and the distance travelled to service them. Homogeneity in workload distribution is computed using the Gini index. Concerning compactness, the measure adopted is the isoperimetric quotient (i.e., the ratio of the polygon area to the area of a circle with same perimeter). Finally, efficiency is obtained by considering the average time taken to travel to any point in the sector from the seed point (which is where the patrol car is hypothetically located).

## 2.5   Conclusions

District design is the problem of grouping elementary units of a given territory into larger districts, according to relevant attributes. Depending on the problem faced, the attributes considered might belong to different contexts, including economical, demographic, geographical, etc. In the last decades, the districting problems have been applied to a broad number of fields. The application of this family of problems to the policing context has given rise to the police districting problem.

In this chapter, a comprehensive review of the police districting problem is given. Initially, a general definition of the problem is provided. Next, the literature on the subject is analyzed in terms of attributes and methodology. Then an annotated bibliography is presented, where the most salient points of each contribution are summarized.

The results of the analysis show that the models proposed in the literature mostly differ on the definitions adopted for the most relevant attributes. In fact, it can be observed a great variability in terms of how sector workload is computed, or on which geometric and topological characteristics should be considered. Also, there is no common agreement on how homogeneity among sectors should be measured. It is the authors' opinion that the research community should work toward a standard definition of the police districting problem. This would allow us to focus most the efforts on a single model, similar to what happened in other areas, such as location analysis or vehicle routing. In particular, it would permit to take steps toward the definition of exact solution approaches for the police districting problem.

**Acknowledgements**  The research of Liberatore and Vitoriano has been supported by the Government of Spain, grant MTM2015-65803-R, and by the Government of Madrid, grant S2013/ICE-2845. All financial supports are gratefully acknowledged. The information and views set out in this paper are those of the author(s) and do not necessarily reflect the official opinion of the financial supporters and of the affiliation institutions.

# References

1. Benveniste, R.: Solving the combined zoning and location problem for several emergency units. J. Oper. Res. Soc. **36**(5), 433–450 (1985)
2. Bodily, S.: Police sector design incorporating preferences of interest groups for equality and efficiency. Manag. Sci. **24**(12), 1301–1313 (1978)
3. Bruce, C.: Districting and resource allocation: a question of balance. Geogr. Public Saf. **1**(4), 1–3 (2009)
4. Bucarey, V., Ordóñez, F., Bassaletti, E.: Shape and balance in police districting. In: Eiselt, H.A., Marianov, V. (eds.) Applications of Location Analysis. International Series in Operations Research and Management Science, vol. 232, pp. 329–347. Springer, Cham (2015)
5. Camacho-Collados, M., Liberatore, F.: A decision support system for predictive police patrolling. Decis. Support. Syst. **75**, 25–37 (2015)

6. Camacho-Collados, M., Liberatore, F., Angulo, J.M.: A multi-criteria police districting problem for the efficient and effective design of patrol sector. Eur. J. Oper. Res. **246**(2), 674–684 (2015)
7. Chaiken, J., Dormont, P.: A patrol car allocation model: background. Manag. Sci. **24**(12), 1280–1290 (1978)
8. Chaiken, J., Dormont, P.: A patrol car allocation model: capabilities and algorithms. Manag. Sci. **24**(12), 1291–1300 (1978)
9. Cheung, C.Y., Yoon, H., Chow, A.H.: Optimization of police facility deployment with a case study in Greater London Area. J. Facil. Manag. **13**(3), 229–243 (2015)
10. Chow, A.H., Cheung, C., Yoon, H.: Optimization of police facility locationing. Transp. Res. Rec. J. Transp. Res. Board **2528**, 60–68 (2015)
11. Curtin, K.M., Qui, F., Hayslett-McCall, K., Bray, T.M.: Integrating GIS and maximal covering models to determine optimal police patrol areas. In: Wang, F. (ed.) Geographic Information Systems and Crime Analysis, chap. 13, pp. 214–235. IGI Global, Hershey (2005)
12. Curtin, K., Hayslett-McCall, K., Qiu, F.: Determining optimal police patrol areas with maximal covering and backup covering location models. Netw. Spat. Econ. **10**(1), 125–145 (2010)
13. D'Amico, S., Wang, S., Batta, R., Rump, C.: A simulated annealing approach to police district design. Comput. Oper. Res. **29**(6), 667–684 (2002)
14. Douglass, J.: Tactical deployment: the next great paradigm shift in law enforcement? Geogr. Public Saf. **1**(4), 6–7 (2009)
15. Grossman, L., Thompson, M., Kluger, J., Park, A., Walsh, B., Suddath, C., Dodds, E., Webley, K., Rawlings, N., Sun, F., Brock-Abraham, C., Carbone, N.: The 50 best inventions. Time Mag. **178**(21), 55–82 (2011)
16. Hess, S., Weaver, J., Siegfeldt, H., Whelan, J., Zitlau, P.: Non-partisan political redistricting by computer. Oper. Res. **13**(6), 998–1008 (1965)
17. Kalcsics, J., Schröeder, M.: Towards a unified territorial design approach—applications, algorithms and GIS integration. TOP **13**(1), 1–74 (2005)
18. Kistler, A.: Tucson police officers redraw division boundaries to balance their workload. Geogr. Public Saf. **1**(4), 3–5 (2009)
19. Konwitz, J.: Cartography in France: 1660-1848. University of Chicago Press, Chicago (1987)
20. Liberatore, F., Camacho-Collados, M.: A comparison of local search methods for the multicriteria police districting problem on graph. Math. Probl. Eng. **2016**, 3690474 (2016)
21. Mayer, A.: Geospatial technology helps east orange crack down on crime. Geogr. Public Saf. **1**(4), 8–10 (2009)
22. Mielke, P.: Using ESRI's districting tool in policing. Geogr. Public Saf. **1**(4), 10–13 (2009)
23. Mitchell, P.: Optimal selection of police patrol beats. J. Crim. Law Criminol. Police Sci. **63**(4), 577–584 (1972)
24. Perry, W., McInnis, B., Price, C., Smith, S., Hollywood, J.: Predictive Policing: The Role of Crime Forecasting in Law Enforcement Operations. RAND Corporation, Santa Monica (2013). Technical report
25. Piyadasun, T., Kalansuriya, B., Gangananda, M., Malshan, M., Bandara, H.D., Marru, S.: Rationalizing police patrol beats using heuristic-based clustering. In: 2017 Moratuwa Engineering Research Conference (MERCon), pp. 431–436. IEEE, Moratuwa (2017)
26. Sarac, A., Batta, R., Bhadury, J., Rump, C.: Reconfiguring police reporting districts in the city of Buffalo. OR Insight **12**(3), 16–24 (1999)
27. Scalisi, N., Beres, J., Sharpe, S., Wilson, R., Brown, T., Whitworth, A. (eds.) Geography and Public Safety Bulletin, vol. 1(4). U.S. Department of Justice, Washington (2009)
28. Wang, F.: Why police and policing need GIS: an overview. Ann. GIS **18**(3), 159–171 (2012)
29. Zhang, Y., Brown, D.: Police patrol districting method and simulation evaluation using agent-based model and GIS. Secur. Inform. **2**(7), 1–13 (2013)
30. Zhang, Y., Brown, D.: Simulation optimization of police patrol district design using an adjusted simulated annealing approach. In: Proceedings of the Symposium on Theory of Modeling and Simulation-DEVS Integrative, article no. 18. Society for Computer Simulation International, San Diego (2014)

31. Zhang, Y., Brown, D.: Simulation optimization of police patrol districting plans using response surfaces. Simulation **90**(6), 687–705 (2014)
32. Zhang, Y., Huddleston, S.H., Brown, D.E., Learmonth, G.P.: A comparison of evaluation methods for police patrol district designs. In: Pasupathy, R., Kim, S.H., Tolk, A., Hill, R., Kuhl, M.E. (eds.) Proceedings of the 2013 Winter Simulation Conference, pp. 2532–2543. IEEE, Piscataway (2013)

# Chapter 3
# A Review of Districting Problems in Health Care

Seda Yanık and Burcin Bozkaya

**Abstract** In this chapter, we review the districting literature in the health care domain. Our goal is to provide the reader with the most relevant studies in the literature as well as a direction for future research. We classify the health care districting problems into three main areas: home care services, primary and secondary health care services, and emergency health care services. We first identify the special characteristics of these different areas. Then we present the modeling approaches, assumptions and solution methods for each of them. In general, we find that certain aspects and dimensions of health care service delivery, which we highlight in this chapter, lend themselves better to the design and implementation of districting-based approaches. As such, we limit our review mostly to studies that include traditional districting models and formulations as well as solution approaches. In closing, we discuss some gaps in the literature and provide directions for future areas of research.

## 3.1 Introduction

Health care service operations management and health care management science have long been active areas of research. Generally speaking, these areas address both the high-level logistics of health care service delivery, that is the planning, structuring, and enabling of the facilities and resources rendering health care services, as well as day-to-day operational planning and scheduling of health care service units and individual resources. A common goal in all of the literature in

S. Yanık
Department of Industrial Engineering, Istanbul Technical University, Istanbul, Turkey
e-mail: sedayanik@itu.edu.tr

B. Bozkaya (✉)
Sabanci School of Management, Sabanci University, Istanbul, Turkey
e-mail: bbozkaya@sabanciuniv.edu

© Springer Nature Switzerland AG 2020
R. Z. Ríos-Mercado (ed.), *Optimal Districting and Territory Design*, International
Series in Operations Research & Management Science 284,
https://doi.org/10.1007/978-3-030-34312-5_3

this domain is increasing the overall efficiency and effectiveness of systems while targeting equitable access to and delivery of health care services by and for the citizen population.

Our goal in this chapter is to review districting-based modeling and solution approaches used in the literature on health care management science. *Districting*, also known or commonly referred to as re-districting, territory design and territory alignment, is the process of dividing a geographical region into smaller areas, which are districts, that represent units of service delivery and are typically collections of smaller sub-areas known as basic units. Districting models and solution approaches find a large variety of applications in such areas as political districting, salesman districting, school districting, and police districting, to name a few. Our focus in this chapter is on the districting-based models along with various relevant criteria within the context of health care service delivery and logistics problems.

As we review the relevant literature, we find three main areas where the underlying health care planning problem is viewed and hence approached as a districting problem: home care services, primary and secondary health care, and emergency care services. Home care services mainly involve caring of elderly people at home who are unable or unfit to travel to fixed medical facilities such as clinics or hospitals. Due to the type of service rendered and the corresponding demand, a natural modeling approach is to group and assign patients or the areas where they reside to districts or "centers" from which services are coordinated. There are various criteria relevant to doing this, such as transportation time, visit time, workload balance and workload limit for health care service personnel, and continuity of service. We present and discuss these in Sect. 3.2.

The second area where districting models and approaches are relevant is the planning of primary and secondary health care. Primary care refers to the first point of contact made by a patient with usually a general practitioner (GP) or family physician (FP). After this first contact, the patient is likely referred to a different facility (e.g., hospitals, clinics) with more capabilities, specialties, expertise, etc. depending on the severity of his or her condition. Here, we find that a significant portion of the relevant studies focuses on the high-level planning that is needed to establish this hierarchical network of fixed facilities (hospitals, GPs and FPs, clinics) and assigning population centers or units to these facilities.

The third area is emergency health care operations, where health care services must be quickly deployed and rendered due to the nature of the emergency, such as a life-threatening situation for an individual or a disaster scenario where large populations are affected. In such cases, a districting approach might also be implemented to identify areas and assign response teams to them that will be dispatched in the case of emergency. The problem, in this case, has a more dynamic nature as the response time must be quick to ensure survival and the time and location of the next request for emergency response are usually treated with a stochastic variable.

In our review, we find that the first two of these three areas more readily lend themselves to optimization models where districts or service areas are created subject to constraints that represent relevant context-related districting criteria. In other words, districts or service areas are created around "centers" at different levels of service hierarchy such as hospitals serving large areas vs. family physicians or GPs serving smaller areas or neighborhoods. As we will see in Sects. 3.2 and 3.3, these models are either in the form of traditional districting formulations where smaller areas are grouped into districts, or location-allocation formulations where district "centers" are located and demand sets representing potential patients are allocated to them. In the emergency health care planning literature, however, we see less of traditional districting formulations and related solution approaches, but more of maximal coverage and maximal survival models with stochastic demand, and in many cases simulation- or queuing-based solution approaches. This is clearly due to the dynamic nature of the problem where the nearest health care service resources demanded (e.g., ambulances) may be in use or limited in number or capacity to respond to some incident, and other units might be deployed from different locations. Hence, a fixed district-based solution approach may not always be suitable. As the main topic of this chapter is districting in health care, we focus mainly on studies with districting, location-allocation, or coverage-based approaches, and choose to highlight only a small selected subset of such studies.

The rest of this chapter is organized into three main sections focusing on three main applications of districting in healthcare: home health care services (Sect. 3.2), primary and secondary health care services (Sect. 3.3), and emergency healthcare services (Sect. 3.4). In each of these three sections, we first discuss the special characteristics of the problem and the most important and relevant districting criteria, then present alternative model formulations, and finally review the literature along with future research directions. In the last section (Sect. 3.5), we present the overall concluding remarks of our study.

## 3.2   Districting of Home Health Care Services

Public health care services need to use effective methods in order to provide the required level of service to the patients with a minimum cost. In today's world, it is known that the elderly population is continuously increasing around the world. This means that the demand for routine and chronic health care services is increasing steadily. In chronic cases, handling the patients at their home environment becomes more efficient. In addition, most of the population in the world today lives in cities. This makes home health care a more efficient option in urban areas. Since the population is densely located in urban areas, the human resources in medical services are more readily accessible. Thus, home health care is becoming a widely used method to service routine patients or elderly people especially in urban areas.

Planning home health care services has many dimensions and associated problems. Two such noteworthy problems involve the districting (or zoning) of the demand points at a tactical or strategic level and the routing and scheduling of the staff at an operational level. The districting problem involves strategic (or tactical) decision making, which aims to identify the service areas that include demand points (i.e., patients) to be served by health care service givers (e.g., nurses). The output

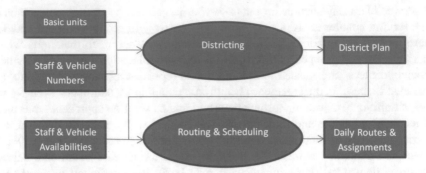

**Fig. 3.1** An overview of the multi-level planning of the home health care services

of districting is the collection of service areas, which is named as the district plan. There are specific features or criteria, which make a district plan a good one. The most common of these are *compactness* and *workload balance* of the districts. The compactness of the districts mainly affects the transportation within the district. Workload balance on the other hand is related to ensuring equity among the staff for fairness and to use the capacity effectively. Routing and scheduling problems are lower level operational problems which specify in what sequence and by whom the daily visits of the demand points (i.e., patients) will be conducted. Based on the district plan generated as a result of the districting study, routing and scheduling plans can be generated on an operational basis. The interplay between these two problems is depicted in Fig. 3.1.

In the following subsections, we examine the home health care districting problem. First, we discuss the special characteristics of home health care services districting. Then, we present the model formulations and finally we review the studies in the literature in this domain.

## 3.2.1 Special Characteristics and Criteria

While partitioning an area into smaller service areas, decision makers or planners must specify the districting criteria used with respect to the operational requirements of home health care services. Similar to the general districting problem, **compactness** is a common criterion used in home health care in order to ensure that the transportation time is small within the districts. Compactness is a criterion that typically refers to how regularly shaped districts are (e.g., similar to squares or circles), and compactness measures can be formulated in many different ways such as those using the distances limits [5, 28] or routing cost approximations [4] or minimizing the maximum distances [5].

In home health care operations, the means of transportation can be various. The health care staff may move on foot, by public transit or by a fleet of vehicles. In some studies, this problem characteristic is addressed in the mathematical models by special measures such as **mobility** [6] or introducing staff sets related to the transportation means they use [28].

The operations of home health care services can be divided into two main time spanning processes: (1) **transportation time** and (2) **visit time**. Thus, most of the studies differentiate between these two types of time-consuming processes. Then, they can be used separately while dealing with the **workload balance** criterion, which refers to how equitable the workload is distributed among health care service providers and is also common in districting studies [16, 18].

Another issue in modeling of the health care districting problem is the specification of the workload. The number of patients is an indicator of the workload. However, a visit to one patient may be significantly different to another patient in terms of workload. Thus, a fine-tuning step may be introduced by categorizing the workloads of patients including a workload balance criterion that considers patient categories. Some examples of categorizing the patients are in terms of age, economic activity, etc. This districting criterion is also known as **population homogeneity** [18].

In addition to the workload balance criterion, **workload limits** can also be introduced in the home health care districting problems. For example, when there is a specific team of staff to service an area, the district workload needs to be compatible with the workload limit of the members of the staff [28].

A criterion which is used frequently in health care is the **continuity of care**, which means a specific staff or a limited number of staff is assigned to visit a particular patient during the episode of care. In health care studies, it is shown that continuity of care increases patient satisfaction, and decreases hospitalization and emergency room visits [4]. Continuity of care especially has positive effects for patients with chronic conditions which is the main type of patients of home health care [4].

Since in most countries, home health care is a public service, **co-extensiveness with current local authorities** (i.e., compatibility) may be required because the regulations of the local authorities may be different. Thus, one of the commonly used criteria in home health care districting forces the basic units of each district to belong to the same administrative unit, in other words to respect the administrative boundaries. This has also been named as compatibility [5, 6].

Some other criteria common to general districting problems are also respected in home health care districting. These are, namely, the **integrality** (i.e., indivisibility of the basic units) and the **contiguity** (i.e., connectivity) of the districts.

## 3.2.2  Model Formulations

There are two main modeling approaches in the home health care districting literature: (1) location-allocation and (2) set partitioning. The choice of the mathematical formulation also drives the solution techniques to be used. Each approach has its advantages and disadvantages. Tackling certain side constraints, such as contiguity, is difficult in location-allocation formulations. In set partitioning models, a set of feasible districts are obtained in the first stage, and then the optimal district plan is found. By the flexibility of the two-stage nature of this approach, many different problem components can be handled. However, as the number of basic units increases, the number of feasible districts increases tremendously. Thus, the computational complexity of the problem increases exponentially as the problem size increases.

First, we introduce a location-allocation formulation in Eqs. (3.1a)–(3.1f). In location-allocation models, the number of districts is generally predefined. However, we may either have a predefined set of district centers or have a potential set of district centers which is the same as the set of basic units. We present the latter one here since it is a generalization of the first one.

Let set $I = \{1, \ldots, n\}$ denote the set of basic units, which are also used as potential district centers. The demand of basic unit $i$ is denoted by $d_i$ and the cost (e.g., distance) between the centroid of basic unit $i$ and basic unit $j$ by $c_{ij}$. Let $x_{ij}$ be a binary decision variable indicating whether basic unit $i$ is included in a district centered at basic unit $j$. Workload equity among districts is formulated using a target level of workload, $\bar{d}$ with a percentage tolerance of $\tau$. The target level of workload is commonly specified as the average workload, which is $\bar{d} = \sum_{i=1}^{n} d_i / m$, where $m$ is the number of districts to be generated.

$$\text{minimize} \quad \sum_{i=1}^{n} \sum_{j=1}^{n} c_{ij} x_{ij} \tag{3.1a}$$

$$\text{subject to} \quad \bar{d}(1 - \tau) \le \sum_{i=1}^{n} d_i x_{ij} \le \bar{d}(1 + \tau) \qquad j = 1, \ldots, n \tag{3.1b}$$

$$\sum_{j=1}^{n} x_{ij} = 1 \qquad\qquad\qquad i = 1, \ldots, n \tag{3.1c}$$

$$x_{ij} \le x_{jj} \qquad\qquad\qquad i, j = 1, \ldots, n \tag{3.1d}$$

$$\sum_{j=1}^{n} x_{jj} = m \tag{3.1e}$$

$$x_{ij} \in \{0, 1\} \qquad\qquad\qquad i, j = 1, \ldots, n \tag{3.1f}$$

Here (3.1a) is the cost minimization objective which also ensures the compactness of the districts to be formed. Constraints (3.1b) guarantee that the workload equity among districts by not allowing workloads to be less than or greater than a certain percentage of the target workload. Constraints (3.1c) together with (3.1d) ensure that each basic unit is assigned to one and only one district, and constraint (3.1e) ensures that only $m$ districts are formed.

In some model formulations, workload equity can be exchanged as an objective instead of the compactness objective. Workload equity is formulated as a min-max equation as in (3.2) together with (3.3) below:

$$\text{minimize} \max_{j=1,\ldots,n:x_{jj}=1} |D_j - \bar{d}| \tag{3.2}$$

$$D_j = \sum_i d_i x_{ij}, \text{ for } j = 1, \ldots, n \tag{3.3}$$

where $D_j$ is introduced as a decision variable denoting the workload of district $j$.

In this case, compactness is included as a constraint in the model introducing a distance bound in two ways: (1) between the basic unit and the district center or (2) between two basic units to be assigned to the same district. A formulation of the first one is presented in (3.4) and the latter is given in (3.5).

$$c_{ij} x_{ij} \leq c_{\max}, \text{ for } i, j = 1, \ldots, n \tag{3.4}$$

where $c_{\max}$ is a bound on the distance between the basic unit and the district center.

$$x_{ij} + x_{kj} \leq 1, \text{ for } i, k \in Z, j = 1, \ldots, n \tag{3.5}$$

where set $Z$ includes basic unit pairs which have a greater pairwise distance than the bound.

A similar formulation can also be employed to ensure the compatibility which forces the basic units of each district to belong to the same administrative unit as in (3.6).

$$x_{ij} + x_{kj} \leq 1, \text{ for } i, k \in T, j = 1, \ldots, n \tag{3.6}$$

where set $T$ includes basic unit pairs that do not belong to the same administrative unit.

The second type of mathematical programming approach used for districting in the literature is set partitioning models. The first stage of this approach uses an auxiliary method to generate feasible districts. In general, at this stage various conditions such as connectivity, workload equity, etc. are evaluated and the feasible districts are identified as the ones that meet the imposed conditions. After this stage, the set partitioning model is formulated to seek the optimal district plan with respect to some cost minimization function. Equations (3.7a)–(3.7d) form a typical set partitioning model. The notation of this model is as follows: $J$ is the set of all feasible districts, $y_j$ is a binary decision variable representing whether feasible district $j$ is included in the optimal district plan or not, $c_j$ is the cost of including district $j$ in a district plan, and $\alpha_{ij}$ is a parameter that shows whether basic unit $i$ is an area in the feasible district $j$.

$$\text{minimize} \quad \sum_{j \in J} c_j y_j \tag{3.7a}$$

$$\text{subject to} \quad \sum_{j \in J} \alpha_{ij} y_j = 1 \qquad i = 1, \dots, n \tag{3.7b}$$

$$\sum_{j \in J} y_j = m \tag{3.7c}$$

$$y_j \in \{0, 1\} \qquad j \in J \tag{3.7d}$$

### 3.2.3   Literature Review and Future Research Directions

Districting of home health care services identifies service regions for the care givers. District plans help with home health care planning and are used commonly together with routing and scheduling tools. As a result of these planning efforts, cost improvements in terms of transportation costs together with visit costs and improvements in the level of health care service provided can be expected. The interested reader may refer to the review studies of home health care services from the perspective of operations research and logistics planning by Emiliano et al. [10], and Milburn [33].

There have been many practical and academic studies related to home health care planning. Here we make a review of the home health care districting studies over the last 15 years. In Table 3.1, we present an overview of criteria, models, and solution methods used in the home health care districting literature.

Blais et al. [6] obtain a near-optimal district plan for a home health care setting in Ontario, Canada using multiple criteria. Together with other well-known criteria such as workload equity, connectivity, etc., they specify a mobility criterion. This criterion captures staff mobility using public transit and walking, which could be possible in a home health care problem. To deal with this problem, they use a multi-criteria clustering method instead of mathematical programming approaches. To find the near-optimal clusters, Tabu search heuristic is employed.

Hertz and Lahrichi [18] divide the workload of the home health care staff into three components: travel load, visit load, and the case load. Case load is specified by the number of patients assigned to different patient categories which correspond to different service requirements. Instead of basic units, they explicitly use locations of patients for districting. Then the patient assignment model is formulated to balance the three workload components in a weighted sum objective function. The solution of the model is sought using a Tabu search algorithm.

Bennett [4] proposes a set-covering model. District workload is used as a constraint to generate feasible districts, and the costs of the feasible districts are defined as an approximation of the routing costs within each district. An initial clustering heuristic and local search improvement method are used to obtain and

**Table 3.1** Home health care districting studies from the literature

| | Criteria (O:objective—C:constraint) | | | | | | | | | | | | Model | | | Solution method |
|---|---|---|---|---|---|---|---|---|---|---|---|---|---|---|---|---|
| Author | Costs | Mobility/accessibility | Workload equity | Patient homogeneity | Visit equity | Travel equity | Integrality | Compactness | Compatibility | Connectivity/contiguous | Time windows | Capacity | Location-allocation | Set partitioning | Clustering | |
| Blais et al. [6] | | O | O | | | | C | | C | C | | | | | ✓ | Tabu search heuristic |
| Hertz and Lahrichi [18] | | | | O | O | O | | | | | | | ✓ | | | Tabu search heuristic |
| Bennett [4] | O | | C | | | | | C | | C | | | | ✓ | | Optimization-based heuristic |
| Benzarti et al. [5] | | | O | | | | C | C | C | | | | ✓ | | | Optimal with gaps |
| Benzarti et al. [5] | | | C | | | | C | O | C | | | | ✓ | | | Optimal with gaps |
| Gutiérrez and Vidal [16] | | | O | | | O | C | | | | | | ✓ | | | Lexicographic algorithm |
| Lin et al. [28] | O | C | | | | | C | C | | | C | C | ✓ | | | Greedy heuristic |

improve the initial feasible districts. Then the set partitioning model is solved using ideas from column generation and heuristic local search methods. First, the linear relaxation of the set partitioning model is solved. Then, a neighborhood search method is used to improve columns to add to the set.

Benzarti et al. [5] present two location-allocation type districting models by switching the most common criteria, namely, compactness and workload equity to be in the first model objective and constraint, respectively, and vice versa in the second model. They also include the implications of different patient requirements on the workload. They implicitly consider the ease of access within the districts as a compatibility criterion. Then the solution of the two models under four different scenarios is obtained and duality between the two types of models is examined.

Gutiérrez and Vidal [16] provide a lexicographic multi-objective model formulation of a districting problem in a rapid-growing city. The two objective function components are the travel workload and total workload equity. They solve the multi-objective model and evaluate the efficient frontier to identify trade-offs.

Lin et al. [28] propose a home health care related districting problem, which is meal distribution service districting. They use a location-allocation modeling approach with additional time windows, maximal travel duration, and capacity issues. They solve the problem using a greedy heuristic method. Then, a case study is investigated and sensitivity analysis on some key parameters is conducted.

The above literature review summarizes the conducted studies that are formulated with deterministic parameters. We also point out, as we reviewed in Sect. 3.2.1, that many differences exist with respect to modeling approaches and the districting criteria used in these studies. Clearly, the research community has not reached a consensus on which criteria to consider and how to handle them (objective function or constraint), which means future researchers will have to face trade-offs between dealing with more complex yet more flexible models and more simplistic models with hard rules. One thing researchers have agreed upon, however, is the use of heuristic approaches to solve the problems. The home health care problem does also involve uncertainty elements and stochastic parameters such as visit times and travel times. Thus, future research directions may include the investigation of stochastic/robust optimization models in the context of home healthcare services.

Moreover, the continuity of care is an important measure in health care services, which has been considered in many practical applications. Yet, this criterion has not been considered in mathematical approaches. As well, the patient case forecasts may also be modeled in dynamic or multi-period home health care districting problems.

The solution methods used to solve both location-allocation and set partitioning models employ heuristics quite commonly. Developing different heuristic solution methods and evaluating different metaheuristic approaches are also additional avenues for future research.

## 3.3   Districting of Primary and Secondary Health Care Services

In most countries, health care services are organized in a hierarchical manner. Commonly, there exists a widespread primary health care service network involving general practitioners. Then the secondary health care services are mainly given in hospitals, which have different levels of amenities and special services. Based on this, the hospitals are categorized into different types such as community hospital, research hospital, etc. The planning of the whole health care system in a certain area involves decisions such as locating the health care service points, specifying which area each point will serve, setting the capacities, allocating resources, scheduling staff, etc. Effective planning of the health care services will help to minimize costs, improve capacity utilization, increase patient satisfaction due to service level, result in accessibility improvements, and ensure equity within society in terms of accessibility to health care services.

One of the most important planning dimensions of the health care services is the designing of the health care service regions that is the districting of health care services. Below we discuss first the special characteristics of health care service districting, then present different model formulations from the literature, and finally review the related literature and provide future research directions for the districting studies related to health care services.

### 3.3.1   Special Characteristics and Criteria

In health care service districting, parallel to the districting models in other domains, spatial **compactness** is one of the major criteria to be honored in the district plan. Spatial compactness may be employed as the total cost of the system, but another important implication of spatial compactness in the setting of health care service districting is that it affects the patients' travel time. In health care service districting, transportation is not maintained by the service giver, and the users of the system commonly reach the centers using the available transportation means. Thus, **accessibility** of patients can be represented within the compactness measure or a separate constraint may also be included in the model. It may be specified in terms of travel distance or travel time or both together [25].

In a health care system, there exist facility-related costs, thus in some studies the facility costs are also considered in the cost formulation. **Facility costs** may include fixed and variable cost of the service capacity. Some additional costs are also identified in the literature due to the mismatch of the facility capacity and the demand. Those include penalty cost of not meeting demand at a location, transportation cost of rerouting of patients, and transport cost of lost or backordered demand [29]. Sometimes, instead of a penalty cost for mismatched demand, a **capacity-demand match** criterion is explicitly included as a constraint.

In health care service districting, **service quality** is another major criterion since the system is designed for the general welfare of the public and to achieve an increase in the patient satisfaction levels. A minimum service level can be imposed as a constraint [29]. Service quality can also be defined in multiple dimensions of which the most common ones are related to the accessibility of patients to the health centers and the performance related to medical procedures. The latter have been explicitly included as separate criteria in some studies. Another health care performance criterion used in the literature is the **variety of medical procedures offered**. Using this criterion, it is aimed to ensure that each district has at least one basic unit with a large variety in the offered medical procedures, so that patients would not have to travel to other districts for many procedures [38]. A more comprehensive criterion used is an aggregate measure named as **health center attractiveness**. Jia et al. [25] formulate this criterion using the following parameters: the number of key special departments, the number of physicians per thousand residents, and the number of beds per thousand residents.

Similar to the home health care districting problems, **population homogeneity** is a criterion used to generate district plans. Population homogeneity may be defined in different terms. Some examples of these terms for achieving population homogeneity in health care service districting are (1) minimizing the deviation of the number of inhabitants in each district from the target average number of inhabitants [9, 38], (2) minimizing the deviation of the value spent on health care of each district from the average spending [9], (3) minimizing the deviation of the size of the elderly population of each district from the average elderly population.

A criterion specified as a constraint in health care service districting is not violating **the range of basic units** included in the districts [9, 38]. This criterion aims to ensure that the number of basic units within each district should be within a defined range.

**Co-extensiveness with current local authorities** (i.e., compatibility) is also an issue considered as a districting criterion in the setting of health care service districting. This criterion forces the basic units of each district to belong to the same administrative unit, in other words to respect the administrative boundaries.

Some other criteria used in health care service districting parallel to the general districting problems are integrity [9, 38], contiguity [9, 38], and absence of complete embedding in another district [38].

### 3.3.2 Model Formulations

Parallel to districting literature, the health care services districting problem has also been modeled using a location-allocation [9, 29, 38] and set partitioning [14] approaches. The model formulations of these two approaches have been provided in Sect. 3.2.2.

A new modeling approach has been presented as a modified $p$-median approach in the literature of health care services districting [25]. A $p$-median model is given in (3.8a)–(3.8e).

$$\text{minimize} \quad \sum_{i=1}^{n} \sum_{j=1}^{m} w_i d_{ij} x_{ij} \tag{3.8a}$$

$$\text{subject to} \quad \sum_{j=1}^{m} x_{ij} = 1 \qquad i = 1, \ldots, n \tag{3.8b}$$

$$x_{ij} - y_j \leq 0 \qquad i = 1, \ldots, n; \ j = 1, \ldots, m \tag{3.8c}$$

$$\sum_{j=1}^{m} y_j = p \tag{3.8d}$$

$$x_{ij}, y_j \in \{0, 1\} \qquad i = 1, \ldots, n; \ j = 1, \ldots, m \tag{3.8e}$$

where $y_j$ is a binary decision variable denoting if a facility (i.e., district center) $j$ is selected, $w_i$ is the weight value of basic unit $i$, $p$ is the number of facilities (districts) to be selected, and $d_{ij}$ is the cost between basic unit $i$ and facility $j$.

A $p$-median model—different from a standard location-allocation model—does not include the capacity limitations. Jia et al. [25] propose to handle the capacity issues of a health care districting problem in the objective function of a $p$-median model. To do this, they define a variable named **shortageRatio** for each facility in each district. The **shortageRatio** is the fraction of the required capacity value (difference between the value of the current capacity and allocated demands) over the current capacity value. Then they include this parameter in the objective function as in (3.9). We note that the inclusion of this additional parameter seems to make the formulation non-linear, yet the authors make no discussion of this non-linearity aspect in their paper nor do they formally include the parameter in their mathematical formulation.

$$\text{minimize} \sum_{j=1}^{m} [(1 + \text{shortageRatio}_j) \cdot \sum_{i=1}^{n} (w_i d_{ij} x_{ij})] \tag{3.9}$$

They also include the compactness constraint in the objective function using a variable named **spatialCost**. This variable is the product of the number of neighboring demanding points with the same facility assignment and a weighting factor $\gamma$.

$$\text{minimize} \sum_{j=1}^{m} [(1 + \text{shortageRatio}_j) \cdot \sum_{i=1}^{n} (w_i d_{ij} x_{ij}) + \text{spatialCost}_j \gamma] \tag{3.10}$$

### 3.3.3   Literature Review and Future Research Directions

In what follows, we report on our review of the health care services districting studies. First, we present a summary of these studies in Table 3.2 with an overview of used criteria, models, and solution methods.

Mahar et al. [29] propose a location-allocation model to determine how many and which of a hospital network's hospitals should be set up to deliver a specialized service and to examine the benefits of pooling these specialized services such as magnetic resonance imaging (MRI), transplants, or neonatal intensive care. Their multi-objective model addresses issues related to capacity-demand match. In the objective function, they include facility costs, the penalty cost of not meeting demand, and the transportation cost of rerouting the patients or backordered demand. They allow fractional demand allocation and solve the model to optimality with gaps. In an application in Indiana, USA, they have examined the effects of customer service level, percent flexible demand, diversion cost, and cost of unmet demand.

Datta et al. [9] deal with a districting problem of primary health care system given by general practitioners. They propose a multi-objective model formulation with the following five objectives: geographical compactness, co-extensiveness with current local authorities, population homogeneity in terms of number of inhabitants, value spent, age which drive morbidity or cost of service. To solve this model, they employ a multiobjective genetic algorithm approach known as NSGA-II. They present a case study in East of England.

Jia et al. [25] use a $p$-median model formulation with an objective including the capacity for the districting problem. The modified $p$-median cost objective is not defined in terms of distance but travel time, which is obtained by taking the road categories and speed limits into account. In the model, they also consider a health care center attractiveness measure, which is calculated using three parameters: the number of key special departments, the number of physicians per thousand residents, and the number of beds per thousand residents. They propose a simulated-annealing based solution methodology to solve the modified $p$-median model. A case study is conducted in Henan Province of China.

Steiner et al. [38] present a location-allocation model for partitioning municipalities into health districts where the number of districts to be generated is a given range instead of a fixed number of districts. They formulate a three-objective model as follows: (1) maximize the homogeneity of inhabitants in the districts, (2) maximize the variety of medical procedures offered in the districts, and (3) minimize inter-district traveling distance. The model formulation involves many constraints such as integrality, contiguity, absence of embedded districts, etc. To solve this model, they employ a multiobjective genetic algorithm approach based on NSGA-II. The health care districting problem of the Parana State of Brazil is investigated in this study.

**Table 3.2** Health care services districting studies from the literature

| Author | Criteria (O: objective—C: constraint) | | | | | | | | | | | | | | | Model | | | Solution method |
|---|---|---|---|---|---|---|---|---|---|---|---|---|---|---|---|---|---|---|---|
| | Facility costs | Mobility/ accessibility | Workload/ size balance | Cost of not meeting demand | Patient homogeneity | Demand fulfillment | Minimum service level | Variety of services | Integrality | Compactness | Compatibility | Connectivity/ contiguous | Absence of embedded districts | District size | Capacity | Location-allocation | Set partitioning | p-Median | |
| Ghiggi et al. [14] | O | | | | | | | | | | | | | | | | | | Backtracking programming |
| Mahar et al. [29] | O | O | | O | | C | C | | | | | | | | C | | ✓ | | Optimal with gaps |
| Datta et al. [9] | | | O | | O | O | C | | | C | O | C | | C | | ✓ | | | Multi-objective GA (NSGA-II) |
| Jia et al. [25] | | O | | | | | | | | C | | | | | | | | ✓ | Simulated Annealing |
| Steiner et al. [38] | | O | O | | O | | | O | C | | | C | C | C | | ✓ | | | Multi-objective GA (NSGA-II) |

The literature on health care service districting is summarized above. As seen, the studies on health care services districting are rather limited. Yet, researchers seem to have agreed, contrary to home health care problems, on modeling these problems with multi-objective approaches. We believe this is an important take-away for future researchers, but at the same time is something that may limit modeling perspectives as the problems at hand continue to evolve thanks to advances in health care logistics and business models. Many extensions can be proposed, however, and some possibilities are as follows: A multi-period or dynamic model would be an interesting research avenue in this area, because making modifications in the health care service system requires high investment costs. Thus, in developing cities, or where the health care demand changes due to increasing elderly population, the costs of the changes required in the system can also be considered in the models. In addition, the hierarchical structure of the health care service system (e.g., primary and secondary) and the interactions between them is also another component to be investigated in the districting models. Finally, satisfaction of the staff of the health care service centers is another issue to be considered as a districting criterion in the models.

## 3.4 Districting of Emergency Health Services

Emergency health care is the third and last area we would like to address in the context of districting and location-allocation based modeling. This area has in fact been subject to an extensive research effort in various modeling and solution approaches, some of which, as mentioned in the Introduction, are outside the scope of this chapter. In what follows, we first describe the literature with the relevant modeling approaches. We then reflect our view on future directions of research.

### 3.4.1 Literature Review

Emergency health care services generally appear in the main context of ambulance or other emergency response vehicle base station location selection, where the related decisions include determining and evaluating coverage and response areas (districts) and vehicle dispatch policies. Because of the highly dynamic and stochastic nature of emergency calls, models that assume the nearest server (i.e., ambulance) will always respond to an incident within its assigned service district are typically not realistic. As a result, many probabilistic approaches with queuing and Markov models are developed. The following literature we discuss mainly highlights such studies with links to districting related issues.

When we consider districting and location-allocation related models only for emergency health care services, the literature goes back as early as the 1960s and 1970s where several studies mark the beginning of location and queuing-based models. Hakimi [17], Toregas et al. [39], Church and ReVelle [8], and Larson [26] are the pioneering authors in this area who have developed the core models that attempt to determine optimum locations for ambulance dispatching facilities with the understanding that each facility location is also associated with a service area or district around it. Hakimi [17] introduces the classical $p$-center problem on a graph where each emergency facility has a coverage radius that corresponds to a "service area" tracing the edges of the graph. Given $p$, the number of locations to determine, the optimal set of locations is the one that minimize the maximum of all coverage radii around the facilities. Each emergency response facility has then all demand or incidents within its coverage radius allocated to it.

Similarly, Toregas et al. [39] consider locating emergency service facilities, but with a set-covering approach. They use a coverage matrix that indicates which candidate location covers which potential demand point within an acceptable distance, and the model seeks to find a minimum number of facility locations that cover all demand points at least once. The selected locations and the covered demand points around them would then constitute the service area for each emergency service facility. Church and ReVelle [8] introduce a closely related formulation, the maximal coverage formulation, where the number of facilities to locate is fixed and the goal is to maximize the coverage of total demand served. Another early study that uses covering-based models is by Hogan and ReVelle [19]. These authors not only propose covering-based formulations to assign potential demand locations to ambulance locations for maximal coverage, thus leading to implicit district definitions, but also consider the case of backup coverage for improved performance on emergency response.

These early studies (and similar others that followed) tackle the demand for emergency health care services from a deterministic point of view, meaning that the potential demand is represented either by the vertices and edges of a graph, or points in two-dimensional space. Larson [26] takes a different angle and considers the problem from a queuing perspective. He proposes a spatially distributed queuing model which he suggests can be useful as an aid to planners in designing emergency response districts. His model involves a continuous-time Markov process with state transition matrix, the state space of which corresponds to the vertices of an $N$-dimensional unit hypercube. A computationally efficient algorithm then is used to tour the vertices of the hypercube to reach a solution for the optimal number of dispatch units.

After these early works on emergency response location modeling, an extensive number of other studies using maximal coverage location formulations (MCLP), spatial queuing models (SQM), and simulation-based approaches have been conducted. Since these are not directly related to districting-based modeling, we leave them outside the scope of this chapter. In fact, even recently, we see a number of studies based on these techniques [12, 13, 20, 22, 34, 37]. We refer the interested reader to comprehensive reviews by Aringhieri et al. [2], Brotcorne et al. [7], Ghobadi et al. [15], Hulshof et al. [21], and Li et al. [27] on these and many other studies.

While the majority of the literature is on the methods and approaches mentioned above, several studies that use the notion of districting in combination with these methods stand out. Iannoni et al. [23] consider the hypercube queuing model and integrate it with a genetic algorithm to determine the locations of emergency response units on a highway network and the primary and secondary response areas (i.e., districts) associated with these locations. Their genetic algorithm chromosome defines a configuration of the ambulance base locations, which is then used to generate a new travel time matrix that is fed into the hypercube model along with the corresponding incident arrival rates. They consider three fitness (objective) functions, including mean response time, fraction of calls not serviced, and the unbalance of servers. The algorithm outputs the final configuration of the ambulance bases along with the resulting primary and secondary service areas around them that are collection of highway segments.

Iannoni et al. [24] further extend the previous study by modifying their hybrid hypercube—genetic algorithm to include the districting decisions in conjunction with location decisions. They implement a two-step version of their algorithm to make location decisions in the first step, followed by districting decisions in the second step, utilizing the algorithm in their earlier work. They apply their algorithm in two case studies on Brazilian highways.

Another related study for locating and districting emergency response vehicles on a network is by Geroliminis et al. [12] who address the three limitations of hypercube models, the most important of which is pre-defining the server locations. They propose both heuristic and exact approaches, which combine the location decisions and coverage variables simultaneously to determine (near-)optimal locations for server bases while considering call response times. Their approach is applied to freeway patrol service optimization; however, the approach and the solution technique are also applicable to emergency health care cases. Recently, other studies have proposed heuristic approaches to handle the complexity in the stochastic nature of the problem and extensive computation times, such as the studies by Toro-Diaz et al. [40] and Enayati et al. [11]. The former study proposes a Tabu search-based heuristic for deciding on ambulance locations with a queuing sub-model, while the latter uses a heuristic approach to generate effective lower bounds to a branch-and-bound implementation.

One other recent study that attempts to locate ambulance stations based on response time and create districts for priority coverage is by Ansari et al. [1]. These authors propose a mixed integer linear programming model to determine station locations and the resulting districts. Different from studies with deterministic demand and service times, they consider uncertainty in travel time as well as ambulance availability, and aim to maximize the coverage level as a result of the location and districting solution. Because of the uncertainty, each incident has a preference list for ambulance locations, leading to a series of response districts that depend on ambulance availability, and the model aims to balance workload among the servers while maintaining contiguity of the first-priority response districts.

While most of the above studies use **response time** as the key performance measure, several recent studies consider a different criterion, **patient survival probability**, in planning emergency health care. For instance, Bandara et al. [3], among others, consider the emergency response problem as a Markov decision process (MDP) to determine optimal dispatching policies, taking into account the degree of call urgency. The resulting policy is the ordered list of ambulances to dispatch, which is determined in terms of patient survivability, depending on the type of incident, not a simple response time threshold. While this study does not have a districting perspective per se, we include this study here to point out the computational results, which are interesting in the sense that they indicate a fixed districting structure may not be suitable. As shown in this study, the shortest response/closest dispatch policy, hinting to the formation of compact districts, is not always optimal in responding to incidents with different priorities.

Similarly, Mayorga et al. [30] propose integrated dispatching and districting policies for emergency medical systems (EMS) and evaluate them from a patient survival probability point of view. They compare different policies through a computer simulation model, evaluate different intra-district and inter-district dispatching policies with respect to patient survivability. They propose a constructive heuristic for building response areas or districts based on expected coverage.

Patel et al. [36] also take into account patient survivability in the special case of cardiac catheterization where they use GIS to evaluate and visualize areas (districts) of accessibility within the province of Alberta, Canada. They consider various modes of transportation in relation to census dissemination areas representing the population centers for possible incidents. They identify zones where the use of one mode would be faster than the others for reaching a treatment facility.

An effective evaluation of the two performance measures mentioned above in emergency medical service dispatching decisions is also provided by McLay and Mayorga [31, 32]. We find other references in the EMS literature with similar approaches to modeling emergency medical response such as MDP, hypercube models, and other queueing-based models. Our conclusion is that while the literature on emergency health care service modeling has started out with "static" or "deterministic" modeling approaches, it has long shifted towards stochastic and queueing-based models, which we choose not to report in this review.

To summarize, our literature review in the context of emergency health care service districting reveals that almost all of the most recent studies make reference to queuing or Markov based models where optimal locations for base ambulance stations are decided and perhaps revised dynamically as necessary, but the associated service or response districts are not fixed areas. This is due to the uncertainty involved in the incident arrival processes and the fact that the nearest server (ambulance) may not always be available or moreover, may not always be the best one to dispatch. As a result, the districts around base locations also tend to be dynamic with many incidents of ambulances dispatched outside their priority service area, entering into service areas of other ambulance base stations. It seems the future models and solution approaches will continue in the same direction.

### 3.4.2 Model Formulation

In this section, we present models from earlier studies [8, 19, 39] in the literature that have attempted to formulate the emergency health response problems in the form of set-covering, location-allocation, or maximal coverage models. More recent models based on MDP, hypercube, or queuing are not included here as they deviate from the classical definition of districting models.

According to Toregas et al. [39], demand for emergency health care services occurs at a finite set of $n$ points indicated by set $I$. The same set of points also represent where potential service providers (e.g., ambulance stations) can be positioned. Assuming a maximum threshold response time of $s$ between a demand point $i$ and a service provider $j$, let $N_i$ denote the set of points within distance $s$ from which emergency service can be provided to demand point $i$. Then let $x_j$ denote the binary decision variable, which takes the value 1 if service provider is established at point $j$. The set-covering based model formulation proposed by Toregas et al. [39] is as follows:

$$\text{minimize} \quad z = \sum_{j=1}^{n} x_j \tag{3.11a}$$

$$\text{subject to} \quad \sum_{j \in N_i} x_j \geq 1 \quad i = 1, \ldots, n \tag{3.11b}$$

$$x_j \in \{0, 1\} \quad j = 1, \ldots, n \tag{3.11c}$$

The objective here is to locate a minimum number of service providers so that all demand points are covered at least once within a distance of $s$. Clearly, this model is not meant to directly generate service districts where all demands located within a district would receive dedicated service from the service provider located in that district, but implicitly achieves a similar effect by forming an $s$-radius service area around each service provider location established. The demand points, however, may have more than one provider within a radius of $s$, hence possibly leading to multiple coverage from various providers nearby.

The concept of multiple coverage has been explicitly treated by Hogan and ReVelle [19] in a bi-objective model as follows:

$$\text{minimize} \quad z_1 = \sum_{j=1}^{n} x_j \tag{3.12a}$$

$$\text{maximize} \quad z_2 = \sum_{i=1}^{n} M_i \tag{3.12b}$$

$$\text{subject to} \quad \sum_{j \in N_i} x_j - M_i \geq 1 \quad i = 1, \ldots, n \tag{3.12c}$$

$$x_j \in \{0, 1\} \qquad j = 1, \ldots, n \qquad (3.12d)$$

$$M_i \in Z \qquad i = 1, \ldots, n \qquad (3.12e)$$

where $M_i$ denotes the number of times demand point $i$ is covered in excess of 1. This is a bi-objective model with a trade-off between the number of service providers located (the lower the better) and the total number of coverages provided to all demand points (the higher the better). We note that with the addition of the second objective regarding multiple coverage, the deviation from a classical districting model is pronounced even further, as a demand point may receive service from a number of provider locations located at varying distances, which means the concept of a "district" is further blurred.

Finally, Church and ReVelle [8] present the maximal coverage model where a fixed number of (emergency) facilities with a given coverage radius are to be located, and the goal is to locate them in such a way that the total demand coverage is maximized. Using the same notation, the formulation for this model is as follows:

$$\text{maximize} \quad z = \sum_{i=1}^{n} a_i y_i \qquad (3.13a)$$

$$\text{subject to} \quad \sum_{j \in N_i} x_j \geq y_i \qquad i = 1, \ldots, n \qquad (3.13b)$$

$$\sum_{j=1}^{n} x_j = p \qquad (3.13c)$$

$$x_j \in \{0, 1\} \qquad j = 1, \ldots, n \qquad (3.13d)$$

$$y_i \in \{0, 1\} \qquad i = 1, \ldots, n \qquad (3.13e)$$

where $y_i$ is the additional decision variable that indicates whether a service demand point $i$ can be covered by any service provider located at $i$ or elsewhere within a response time of $s$; $a_i$ is the demand to be served at point $i$; and $p$ is the total number of emergency facilities to be located. As implied by the objective function, this model further takes into account the "population" to be served, i.e., not just a demand point being "covered." It can also be considered as a complementary model to that of Toregas et al. [39] as there is a clear trade-off between the number of facilities to be located (hence their cost) and the total coverage they provide.

### 3.4.3   Future Research Directions

In our review, we find that a significant portion of the recent and relevant literature on emergency health care service districting is in fact related to queuing-based models in combination with optimization models. As in the case of hypercube queuing models used together with genetic algorithms or Tabu search, there are many complexities in using queuing-based approaches in an optimization framework. As some of the most recent studies suggest, we think that a continued and viable line of future research would be the further exploration of heuristic methods such as adaptive large neighborhood search, swarm optimization, matheuristics, etc. The use of such heuristics might prove to be more efficient, leading to improvements in computational times and hence a better and more exhaustive search over solution alternatives.

We also think that other criteria for emergency health care planning might be relevant. While incident response times and patient survivability rates are of utmost importance for rendering emergency health care service, the cost of these systems cannot be overlooked. In this regard, models that address the cost versus service level trade-off are still in need, as the emergency health care response problem is in essence a queuing problem for scarce resources with high potential impact on saving human life but at the same time very costly.

As the traditional districting approaches do not completely fit in the framework of emergence response modeling, more dynamic versions of districting models in combination with queuing approaches could also be worthwhile exploring. Districting typically makes it much easier and comprehensible for decision makers and system planners in conducting day-to-day operations. To take advantage of this, existing districting models might be adapted to more dynamic or stochastic settings, where district boundaries might be changing over time and districts might have overlapping regions to account for possibilities of multiple coverage. This has been done only very recently [11, 35], and the districting literature is yet to see more of these kinds of variations applied to emergency care response modeling.

Finally, as we live in the age of Big Data and Machine Learning, we believe there is great potential value in making predictions about emergency call incidents, road conditions and travel times, and server utilization. As the EMS systems collect massive amounts of data from day-to-day operations, these datasets may be analyzed further to develop better models for the logistical infrastructure of emergency health care systems. While we recently started seeing some studies in this arena, this is yet an area we believe has not taken off yet.

## 3.5   Conclusion

In this chapter, we took a closer look at districting problems arising in the health care services context as well as districting and location-allocation based models proposed to solve them. Operations management and management science literature addressing health care service management issues and aimed at solving related problems have been around for decades. Here we consider three main categorizations of these problems: (a) home health care services; (b) primary and secondary health care; and (c) emergency health care services. A great variety of such services require systematic effort and logistical organization; hence, they readily lend themselves to districting-based problem formulations.

With home health care services management, we find that as the population gets older, more and more home care services are deployed, towards the aging population who are unable or unfit to travel to fixed medical facilities. In this context, many day-to-day operational activities of health care professionals, such as nurses and doctors, to render these services are organized around service districts. Hence, we see an extensive number of districting-based models and formulations under various criteria for achieving the operational efficiency of these services. We deduct from the reviewed literature that the modeling efforts mainly focus on achieving the equity of workload among nurses/doctors as well as maintaining the geometry related criteria such as connectivity, compactness, and integrality, which in return affect the cost and time of traveling of the home healthcare vehicles. Thus, we can say that the home health care service design problems seem to best fit the traditional districting modeling where typically location-allocation, set-covering, and clustering approaches are used. The employed solution techniques are generally heuristic-based approaches since the NP-hard nature of the model does not allow to find optimal solutions for the real-size problems.

Primary and secondary health care requires organization at a higher level, with various types of medical facilities offering services with varying capacities and specialties serving in different regions. This necessitates district-based modeling at various levels of hierarchy, and our literature survey seems to support it. The literature related to primary and secondary health care design problems deals mainly with accessibility and demand-related issues such as variety of services, demand fulfillment, and service levels. Hence, the modeling approaches used are also traditional districting and facility location techniques such as location-allocation, set partitioning, and p-median, which are typically solved using heuristic approaches as well. We further find that new studies and modeling approaches addressing the health care demand and supply mismatch due to population movements or demography shifts are needed.

The emergency healthcare service aims to achieve mainly a certain response time to reach the incidents. Thus, it is the main criterion used in this domain together with demand fulfillment, patient/region priorities, and balance of the servers. The literature, however, has recently concentrated around spatial queuing models and simulation approaches due to the highly probabilistic nature of emergency care incidents and scarcity of servers (e.g., ambulances). This is clearly an improved line of research over the districting and location-allocation based modeling approaches for the problem developed in the 1970s and 1980s. Lately, heuristic optimization in combination with these queuing-based modeling approaches seems to have picked up trend. Some of the models we review have clear districting implications with the emergency response units still having to operate within well-defined service areas. Yet the link between districting-based formulations and queuing-based models is rather weak due to the dynamic nature of the problems involved and perhaps stochastic or fuzzy districting modeling approaches might offer further contributions in the future.

In conclusion, we note that we expect this line of research on health care operations management to continue in years to come, and that districting-based models and solution approaches will further contribute to the literature. We hope that our study sheds light to some of the recent work on districting and health care services management and lead future researchers to take new angles and generate new ideas in solving modern health care management problems.

# References

1. Ansari, S., McLay, L.A., Mayorga, M.E.: A maximum expected covering problem for district design. Transp. Sci. **51**(1), 376–390 (2015)
2. Aringhieri, R., Bruni, M.E., Khodaparasti, S., van Essen, J.T.: Emergency medical services and beyond: addressing new challenges through a wide literature review. Comput. Oper. Res. **78**, 349–368 (2017)
3. Bandara, D., Mayorga, M.E., McLay, L.A.: Optimal dispatching strategies for emergency vehicles to increase patient survivability. Int. J. Oper. Res. **15**(2), 195–214 (2012)
4. Bennett, A.R.: Home health care logistics planning. Ph.D. Thesis, Georgia Institute of Technology, Atlanta (2009)
5. Benzarti, E., Sahin, E., Dallery, Y.: Operations management applied to home care services: analysis of the districting problem. Decis. Support. Syst. **55**(2), 587–598 (2013)
6. Blais, M., Lapierre, S.D., Laporte, G.: Solving a home-care districting problem in an urban setting. J. Oper. Res. Soc. **54**(11), 1141–1147 (2003)
7. Brotcorne, L., Laporte, G., Semet, F.: Ambulance location and relocation models. Eur. J. Oper. Res. **147**(3), 451–463 (2003)
8. Church, R., ReVelle, C.: The maximal covering location problem. Pap. Reg. Sci. Assoc. **32**(1), 101–118 (1974)
9. Datta, D., Figueira, J.R., Gourtani, A.M., Morton, A.: Optimal administrative geographies: an algorithmic approach. Socio Econ. Plan. Sci. **47**, 247–257 (2013)
10. Emiliano, W., Telhada, J., do Sameiro Carvalho, M.: Home health care logistics planning: a review and framework. Procedia Manuf. **13**, 948–955 (2017)
11. Enayati, S., Ozaltin, O.Y., Mayorga, M.E., Saydam, C.: Ambulance redeployment and dispatching under uncertainty with personnel workload limitations. IISE Trans. **50**(9), 777–788 (2018)
12. Geroliminis, N., Karlaftis, M.G., Skabardonis, A.: A spatial queuing model for the emergency vehicle districting and location problem. Transp. Res. B **43**(7), 798–811 (2009)
13. Geroliminis, N., Kepaptsoglou, K., Karlaftis, M.G.: A hybrid hypercube-genetic algorithm approach for deploying many emergency response mobile units in an urban network. Eur. J. Oper. Res. **210**(2), 287–300 (2011)
14. Ghiggi, C., Puliafito, P.P., Zoppoli, R.: A combinatorial method for health-care districting. In: Cea, J. (ed.) Optimization Techniques: Modeling and Optimization in the Service of Man Part I. Lecture Notes in Computer Science, vol. 40, pp. 116–130. Springer, Berlin (1976)
15. Ghobadi, M., Arakat, J., Tavakkoli-Moghaddam, R.: Hypercube queuing models in emergency service systems: a state-of-the-art review. Sci. Iran. **26**(2), 909–931 (2019)
16. Gutiérrez, E.V., Vidal, C.J.: A home health care districting problem in a rapid-growing city. Ing. Univ. **19**(1), 87–113 (2015)
17. Hakimi, S.: Optimum locations of switching centers and the absolute centers and medians of a graph. Oper. Res. **12**(3), 450–459 (1964)
18. Hertz, A., Lahrichi, N.: A patient assignment algorithm for home care services. J. Oper. Res. Soc. **60**(4), 481–495 (2009)

19. Hogan, K., ReVelle, C.: Concepts and applications of backup coverage. Manag. Sci. **32**(11), 1434–1444 (1986)
20. Hosseini, M.H.M., Ameli, M.S.J.: A bi-objective model for emergency services location-allocation problem with maximum distance constraint. Manag. Sci. Lett. **1**(2), 115–126 (2011)
21. Hulshof, P.J.H., Kortbeek, N., Boucherie, R.J., Hans, E.W., Bakker, P.J.M.: Taxonomic classification of planning decisions in health care: a structured review of the state of the art in OR/MS. Health Syst. **1**(2), 129–175 (2012)
22. Iannoni, A.P., Morabito, R.: A multiple dispatch and partial backup hypercube queuing model to analyze emergency medical systems on highways. Transp. Res. E **43**(6), 755–771 (2007)
23. Iannoni, A.P., Morabito, R., Saydam, C.: A hypercube queueing model embedded into a genetic algorithm for ambulance deployment on highways. Ann. Oper. Res. **157**(1), 207–224 (2008)
24. Iannoni, A.P., Morabito, R., Saydam, C.: An optimization approach for ambulance location and the districting of the response segments on highways. Eur. J. Oper. Res. **195**(2), 528–542 (2009)
25. Jia, T., Tao, H., Qin, K., Wang, Y., Liu, C., Gao, Q.: Selecting the optimal healthcare centers with a modified p-median model: a visual analytic perspective. Int. J. Health Geogr. **13**(42), 1–15 (2014)
26. Larson, R.C.: A hypercube queuing model for facility location and redistricting in urban emergency services. Comput. Oper. Res. **1**(1), 67–95 (1974)
27. Li, X., Zhao, Z., Zhu, X., Wyatt, T.: Covering models and optimization techniques for emergency response facility location and planning: a review. Math. Meth. Oper. Res. **74**(3), 281–310 (2011)
28. Lin, M., Chin, K., Fu, C., Tsui, K.: An effective greedy method for the meals-on-wheels service districting problem. Comput. Ind. Eng. **106**, 1–19 (2017)
29. Mahar, S., Bretthauer, K.M., Salzarulo, P.A.: Locating specialized service capacity in a multi-hospital network. Eur. J. Oper. Res. **212**(3), 596–605 (2011)
30. Mayorga, M.E., Bandara, D., McLay, L.A.: Districting and dispatching policies for emergency medical service systems to improve patient survival. IIE Trans. Healthc. Syst. Eng. **3**(1), 39–56 (2013)
31. McLay, L.A., Mayorga, M.E.: Evaluating the impact of performance goals on dispatching decisions in emergency medical service. IIE Trans. Healthc. Syst. Eng. **1**(3), 185–196 (2011)
32. McLay, L.A., Mayorga, M.E.: A model for optimally dispatching ambulances to emergency calls with classification errors in patient priorities. IIE Trans. **45**(1), 1–24 (2013)
33. Milburn, A.B.: Operations research applications in home healthcare. In: Hall, R. (ed.) Handbook of Healthcare System Scheduling. International Series in Operations Research and Management Science, vol. 168, chap. 11, pp. 281–302. Springer, Boston (2011)
34. Mohammadi, M., Khotbesara, Z., Mirzazadeh, A.: MCLP and SQM models for the emergency vehicle districting and location problem. Decis. Sci. Lett. **3**(4), 479–490 (2014)
35. Nasrollahzadeh, A.A., Khademi, A., Mayorga, M.E.: Real-time ambulance dispatching and relocation. Manuf. Serv. Oper. Manag. **20**(3), 467–480 (2018)
36. Patel, A.B., Waters, N.M., Ghali, W.A.: Determining geographic areas and populations with timely access to cardiac catheterization facilities for acute myocardial infarction care in Alberta, Canada. Int. J. Health Geogr. **6**(47), 1–12 (2007)
37. Peleg, K., Pliskin, J.S.: A geographic information system simulation model of EMS: reducing ambulance response time. Am. J. Emerg. Med. **22**(3), 164–170 (2004)
38. Steiner, M.T.A., Datta, D., Neto, P.J.S., Scarpin, C.T., Figueira, J.R.: Multi-objective optimization in partitioning the healthcare system of Parana State in Brazil. Omega **52**, 53–64 (2015)
39. Toregas, C., Swain, R., ReVelle, C., Bergman, L.: The location of emergency service facilities. Oper. Res. **19**(6), 1363–1373 (1971)
40. Toro-Diaz, H., Mayorga, M.E., McLay, L.A., Rajagopalan, M.A., Saydam, C.: Reducing disparities in large-scale emergency medical service systems. J. Oper. Res. Soc. **66**(7), 1169–1181 (2015)

# Chapter 4
# Computational Geometric Approaches to Equitable Districting: A Survey

Mehdi Behroozi and John Gunnar Carlsson

**Abstract** Dividing a piece of land into sub-regions is a natural problem that belongs to many different domains within the world of operations research. There are many different tools that one can use to solve such problems, such as infinite-dimensional optimization, integer programming, or graph theoretic models. In this chapter, we summarize previous methods for solving districting problems that use computational geometry.

## 4.1 Introduction

Dividing a piece of land into sub-regions is a natural problem that belongs to many different domains within the world of operations research, such as air traffic control, congressional districting, vehicle routing, facility location, urban planning, and supply chain management. Indeed, effective division of geographic territory has been a fundamental societal problem since the times of antiquity [15]:

> Homer, in describing the Phaiakian settlement in Scheria, speaks of a circuit wall for the city.... Implicit in the foundation of new colonies was the notion of equality among the members, exemplified in the division of their prime resource, the land. To achieve this, accurate measurement and equitable division were from the outset essential, even when gods or privileged men were to be honored with larger or better assignments.

Scientifically speaking, one of the major difficulties that problems of this type pose is their intrinsically interdisciplinary nature; in order to determine an optimal partition of a territory, one must combine tools from a variety of disciplines. The purpose of this chapter is to identify and summarize various algorithms for territory

M. Behroozi
Northeastern University, Boston, MA, USA
e-mail: m.behroozi@northeastern.edu

J. G. Carlsson (✉)
University of Southern California, Los Angeles, CA, USA
e-mail: jcarlsso@usc.edu

© Springer Nature Switzerland AG 2020
R. Z. Ríos-Mercado (ed.), *Optimal Districting and Territory Design*, International
Series in Operations Research & Management Science 284,
https://doi.org/10.1007/978-3-030-34312-5_4

districting that use *computational geometry*. This is distinct from, for example, explicitly discrete network-based models, such as graph partitioning and multiway cuts [13]. The key difficulty in such problems is the balancing of *allocation objectives*, such as some measure of workload in the sub-regions or balanced consumption of a resource, with geometric *shape constraints*. As examples of shape conditions, we may require that all sub-regions be compact, convex, simply connected (not having holes), connected, or merely measurable. Because of the extreme breadth of literature on the subject of districting, we will focus this chapter only on approaches to districting problems with the following attributes: the input region is a Euclidean domain $\mathcal{R}$, usually $\mathbb{R}^2$ or a subset thereof (as opposed to a more abstract space, such as a graph). Second, the objective is to divide $\mathcal{R}$ into districts while optimizing an objective function or satisfying a criterion (or both). Third, the problems of interest have some practical application to districting, as opposed to being computationally oriented (e.g., finding a good mesh triangulation for numerical integration).

### 4.1.1 Notational Conventions

The notational conventions for this problem are as follows: the input is a Euclidean region $\mathcal{R}$, possibly with additional information such as a probability density or a set of points. We seek a *partition* of $\mathcal{R}$ into districts $D_1, \ldots, D_n$, that is, a collection of disjoint subsets such that $\bigcup_{i=1}^{n} D_i = \mathcal{R}$ and interior$(D_i) \cap$ interior$(D_j) = \emptyset$. The boundary of a set $S$ is written $\partial S$, the line segment between points $a$ and $b$ is written $\overline{ab}$, and $\| \cdot \|$ denotes the Euclidean norm unless otherwise stated.

### 4.1.2 Applications of Geometric Districting

As mentioned in the introduction, there are many natural operational problems for which districting—from a geometric viewpoint—is a useful tool. One popular application is in air traffic control, as can be found in [7, 53]; here the goal is to balance the workloads within districts, interpreted as the maximum number of airplanes contained within a district at any time. The use of partitioning to balance workloads is a pervasive theme in robotics and vehicle routing as well, as seen in [20, 21, 23, 24, 42, 43]. One recurrent theme in many of these papers is the application of the foundational results in geometric probability theory, the *Beardwood–Halton–Hammersley Theorem* [8], which describes the length of the optimal travelling salesman tour of a set of points sampled from a probability distribution. Specifically, if $\{X_i\}$ is a sequence of independent samples drawn from an absolutely continuous probability distribution $f(\cdot)$ defined on a compact planar region $\mathcal{R}$, then the length $L_N$ of the shortest tour through points $X_1, \ldots, X_N$ satisfies

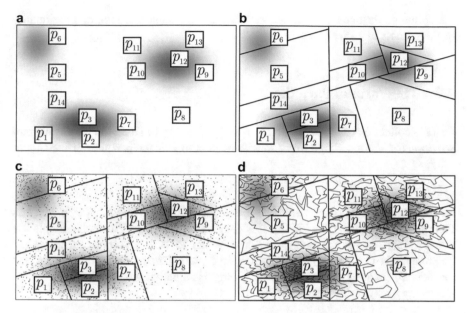

**Fig. 4.1** The diagram represents the stochastic multi-depot vehicle routing problem formulated as a geometric districting problem; we begin with a set of $n = 13$ vehicle "depots" $p_i$ with fixed locations and a probability density $f(\cdot)$ defined on $R$ (**a**), which we then partition into $n$ convex districts (**b**). This partition should be constructed so that, when a large collection of points is sampled independently from $f(\cdot)$ as in (**c**), the $n$ TSP tours of all the points in each sub-region plus the depot point are balanced, on average (**d**). This is guaranteed to happen within $o(\sqrt{N})$ if $\int_{D_i} \sqrt{f(x)}\, dx$ is equal for all districts

$$L_N = \sqrt{N} \int_{\mathcal{R}} \sqrt{f(x)}\, dx + o(\sqrt{N}) \quad \text{with probability 1}$$

as $N \to \infty$. A consequence of this result is the fact that, if one partitions $\mathcal{R}$ into districts $D_1, \ldots, D_n$ such that $\int_{D_i} \sqrt{f(x)}\, dx$ is equal for all $i$, then we are guaranteed that the TSP tours of the samples within each district are within $o(\sqrt{N})$ of each other as $N \to \infty$; see Fig. 4.1.

Districting problems can also be used to solve facility location problems. The seminal paper [39] contains many such examples, as do the more recent contributions [25, 36]. One of the key concepts is the fact that one can use a particular districting scheme—such as the *Voronoi partition* which we will describe in the following section—to reduce the set of candidate placements of a facility to a finite set.

Not surprisingly, computational geometry has had a significant impact on congressional districting (gerrymandering) and school district design [18, 26, 50, 51]. Here the complexity comes from the sheer multitude of factors that influence an optimal decision, such as the number of students per district (of each demographic), the commute distance between students and their assigned school, and the quality of the schools in each district.

## 4.2 The Voronoi Paradigm

Given a collection of landmark points $P = \{p_1, \ldots, p_n\}$ in a region $\mathcal{R}$, the *Voronoi partition* (or Voronoi diagram) of $\mathcal{R}$ with respect to $P$ is the partition defined by

$$D_i = \{x \in \mathcal{R} : \|x - p_i\| \leq \|x - p_j\| \; \forall j\}.$$

Not surprisingly, Voronoi partitions are of fundamental importance in location theory and spatial analysis [4, 40]. It is easy to see that the districts $D_i$ are always convex when $\| \cdot \|$ is the Euclidean norm. When $\mathcal{R} = \mathbb{R}^2$, Voronoi partitions can be computed in $\mathcal{O}(n \log n)$ running time [6, 27, 48]. The boundary segments of the districts $D_i$ (as well as their endpoints) comprise a planar graph, whose dual is the *Delaunay triangulation* of $P$; see Fig. 4.2a, b.

### 4.2.1 Weighted Voronoi Partitions

A *weighted* Voronoi partition is a subdivision of $\mathcal{R}$ into districts $D_i$, taken simultaneously with respect to $P$ and a weight vector $w \in \mathbb{R}^n$ that "discounts" or "surcharges" the distances. Different weighting schemes give rise to districts with different shape properties. In an *additively* weighted Voronoi partition, the districts are defined by

$$D_i = \{x \in \mathcal{R} : \|x - p_i\| - w_i \leq \|x - p_j\| - w_j \; \forall j\};$$

it is easy to see that the boundaries between adjacent sub-regions are hyperbolic arcs because

$$\partial D_i \cap \partial D_j \subset \{x \in \mathcal{R} : \|x - p_i\| - \|x - p_j\| = w_i - w_j\}$$

and a hyperbola is defined as the locus of points $x$ such that the difference in distances to two landmark points $p_i, p_j$ is constant. It is also easy to verify that the districts are connected (and in fact star-convex relative to the landmark points $P$) because if $x \in D_i$, then the line segment $\overline{xp_i}$ also belongs to $D_i$. See Fig. 4.2c.

In a *power* Voronoi partition (often simply called a power diagram), the districts are defined by

$$D_i = \{x \in \mathcal{R} : \|x - p_i\|^2 - w_i \leq \|x - p_j\|^2 - w_j \; \forall j\}.$$

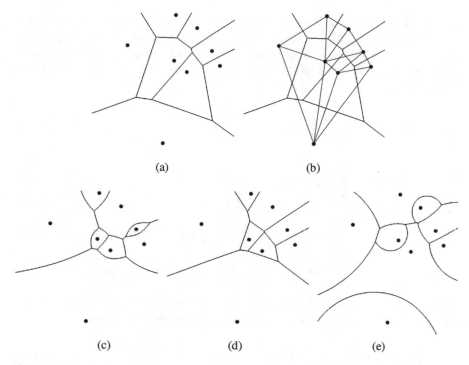

**Fig. 4.2** Voronoi partitions and related structures. (**a**) Voronoi. (**b**) Delaunay triangulation (**c**) Additive Voronoi. (**d**) Power Voronoi. (**e**) Multiplicative Voronoi

It is easy to see that the boundaries between adjacent sub-regions are line segments and consequently that all districts are convex, although the situation can easily arise that $p_i \notin D_i$ for some $i$. See Fig. 4.2d.

In a *multiplicatively* weighted Voronoi diagram, the districts are defined by

$$D_i = \{x \in \mathcal{R} : \|x - p_i\|/w_i \leq \|x - p_j\|^2/w_j \ \forall j\};$$

the *Apollonius circle theorem* says that the boundaries between adjacent sub-regions are circular arcs [38]. See Fig. 4.2e.

## 4.2.2   *Complementary Slackness and Weighted Voronoi Diagrams*

The notion of *complementary slackness* provides an elegant relationship between assignment problems and their counterparts in districting. This is best exemplified in [5], which establishes a bijection between power Voronoi partitions and least-squares assignments of points to landmarks. Suppose for simplicity that $\mathcal{R} = \mathbb{R}^2$

(or indeed any $\mathbb{R}^d$) and consider a set of landmark points $P$ as before, as well as a set of *demand points* $Q = \{q_1, \ldots, q_m\}$ and a *capacity vector* $c \in \mathbb{Z}_+^n$ such that $\sum_{i=1}^n c_i = m$. The least-squares problem of assigning demand points to landmark points, subject to the constraint that $p_i$ is assigned to exactly $c_i$ demand points, is written as

$$\underset{X}{\text{minimize}} \quad \sum_{i=1}^n \sum_{j=1}^m d_{ij}^2 x_{ij} \tag{4.1a}$$

$$\text{subject to} \quad \sum_{j=1}^m x_{ij} = c_i \qquad i = 1, \ldots, n \tag{4.1b}$$

$$\sum_{i=1}^n x_{ij} = 1 \qquad j = 1, \ldots, m \tag{4.1c}$$

$$x_{ij} \in \{0, 1\} \qquad i = 1, \ldots, n; \; j = 1, \ldots, m \tag{4.1d}$$

where $d_{ij} = \|p_i - q_j\|$. Of course, the integer program above has no integrality gap and can hence be relaxed to a linear program, whose dual is given by

$$\underset{\lambda, \nu}{\text{maximize}} \quad \sum_{i=1}^n c_i \lambda_i + \sum_{j=1}^m \nu_j \tag{4.2a}$$

$$\text{subject to} \quad \lambda_i + \nu_j \le d_{ij}^2 \qquad i = 1, \ldots, n; \; j = 1, \ldots, m \tag{4.2b}$$

Note that, for any optimal solution $(\lambda^*, \nu^*)$ to the above, it must also be the case that $(\lambda^* + t, \nu^* - t)$ is also optimal for all scalar $t$. Therefore, we can assume without loss of generality that $\sum_{i=1}^n c_i \lambda_i = 0$ which reduces (4.2) to the piecewise linear concave maximization problem

$$\underset{\lambda}{\text{maximize}} \quad \sum_{j=1}^m \min_i \{d_{ij}^2 - \lambda_i\} \tag{4.3a}$$

$$\text{subject to} \quad \sum_{i=1}^n c_i \lambda_i = 0. \tag{4.3b}$$

Complementary slackness between (4.3) and the linear relaxation of (4.1) establishes the following bijections:

**Theorem 4.1** *Let* $X^*$ *and* $\lambda^*$ *denote optimal primal-dual pairs between between (4.3) and the linear relaxation of (4.1). For* $i \in \{1, \ldots, n\}$, *let* $S_i = \{q_j : x_{ij}^* = 1\}$ *denote the set of demand points assigned to* $i$. *Then the power Voronoi partition with respect to weight vector* $w = \lambda^*$, *i.e.*

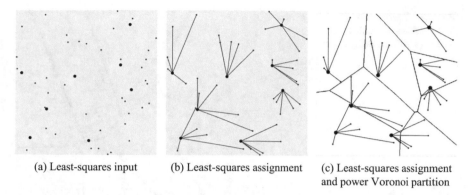

(a) Least-squares input    (b) Least-squares assignment    (c) Least-squares assignment
                                                            and power Voronoi partition

**Fig. 4.3** The equivalence established by Aurenhammer et al. [5]. (**a**) shows an input with $n = 8$ and $m = 80$, (**b**) shows the optimal assignment when $c_i = 5$ for all $i$, and (**c**) shows the power Voronoi partition that induces the optimal assignment

$$D_i = \{x \in \mathcal{R} : \|x - p_i\|^2 - \lambda_i^* \leq \|x - p_j\|^2 - \lambda_j^* \ \forall j\},$$

*satisfies the property that* $S_i \subset D_i$ *for all* $i$.

**Theorem 4.2** *For any weight vector* $w$ *that induces a power Voronoi partition* $D_1, \ldots, D_n$, *let* $S_i = \{q_j : q_j \in D_i\}$, *with ties broken arbitrarily if* $q_j$ *lies on a boundary, and set* $c_i = |S_i|$. *Then the solution* $X$ *obtained by setting* $x_{ij} = 1$ *if* $q_j \in D_i$ *and zero otherwise is optimal for (4.1).*

**Corollary 4.1** *For any* $P$ *and* $Q$ *in general position and any* $c \in \mathbb{Z}_+^n$ *such that* $\sum_{i=1}^n c_i = m$, *there exists a weight vector* $w$ *whose induced power Voronoi partition satisfies* $|D_i \cap Q| = c_i$ *for all* $i$.

Figure 4.3 shows an example of the above results. In Corollary 4.1, the existence of $w$ is guaranteed by setting $w = \lambda^*$.

The preceding results are surprisingly general and have influenced many further studies. For example, almost nothing changes if we consider a *semi-infinite* version of (4.1) in which we replace the point set $Q$ with a probability density $f$, and thereby replace the summation with an integral; the assignment variable $x_{ij}$ now becomes an "assignment function" $\mathcal{I}_i(x)$ that indicates if point $x$ is assigned to landmark point $i$:

$$\underset{\mathcal{I}_1(\cdot), \ldots, \mathcal{I}_n(\cdot)}{\text{minimize}} \quad \sum_{i=1}^n \int_{\mathcal{R}} \|x - p_i\|^2 f(x) \mathcal{I}_i(x) \, dx \tag{4.4a}$$

$$\text{subject to} \quad \int_{\mathcal{R}} f(x) \mathcal{I}_i(x) \, dx = c_i \qquad i = 1, \ldots, n \tag{4.4b}$$

**Fig. 4.4** A power diagram
that partitions the unit square,
with all districts having equal
area

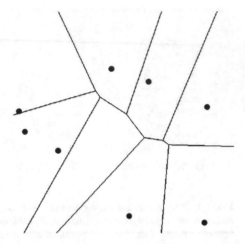

$$\sum_{i=1}^{n} \mathcal{I}_i(x) = 1 \qquad\qquad x \in \mathcal{R} \qquad\qquad (4.4c)$$

$$\mathcal{I}_i(x) \in \{0, 1\} \qquad\qquad i = 1, \dots, n; \ x \in \mathcal{R} \qquad (4.4d)$$

This fact was also noted in [5]; as an example, if we set $f(x) = 1$ everywhere and
$c_i = \text{Area}(\mathcal{R})/n$ for all $i$, then the preceding results guarantee that one can always
find a power Voronoi partition in which all districts have equal area; see Fig. 4.4.
This analysis was taken even further in [17], which describes the family of *all*
weighted clusterings expressible as power diagrams via so-called *gravity polytopes*.

Indeed, the basic principle remains unchanged if one replaces the squared
distance functions $\|x - p_i\|^2$ with other cost functions $\phi(x - p_i)$. For example,
[3] and [24] observe that if one uses the standard Euclidean distance $\phi(z) = \|z\|$
instead, then one obtains an additively weighted Voronoi diagram with the same
properties (Fig. 4.5a). If one substitutes the Manhattan norm $\phi(z) = \|z\|_1$, [23]
shows that the boundary components are line segments in the eight cardinal and
ordinal directions, and also gives extensions for the case where $\phi(z)$ is a distance
metric imposed by polygonal obstacles. Furthermore, Cortés [24] notes that if one
uses $\phi(z) = \log \|z\|$, then one obtains a multiplicatively weighted Voronoi diagram
(Fig. 4.5c) and gives a Jacobi iterative algorithm for finding both diagrams quickly.
With an application towards electoral districting, [18] consider graph-based distance
functions as well as ellipsoidal distance functions, i.e., $\phi_i(z) = z^T M_i z$ (or the
square root thereof) for positive definite $M_i$, which they call *anisotropic Voronoi
diagrams*. The boundary curves are either conic sections or quartic curves depending
on whether or not the square root is used.

The paper [21] describes a related problem to (4.4) where one is given a
probability density $f$, but the constraints on the masses $\int_{\mathcal{R}} f(x)\mathcal{I}_i(x)\,dx = c_i$ are
dropped. Rather, the objective is to minimize the *maximum* cost over all districts:

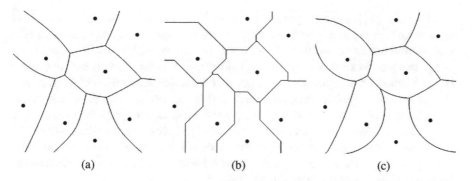

**Fig. 4.5** Equal-area weighted Voronoi diagrams with varying cost functions; in (**b**), the boundary components are line segments in the eight cardinal and ordinal directions. (**a**) Euclidean distance. (**b**) Manhattan distance. (**c**) Logarithmic distance

$$\underset{\mathcal{I}_1(\cdot),\dots,\mathcal{I}_n(\cdot)}{\text{minimize}} \quad \underset{i}{\max} \int_{\mathcal{R}} \phi_i(x)f(x)\mathcal{I}_i(x)\,dx$$

$$\text{subject to} \quad \sum_{i=1}^{n} \mathcal{I}_i(x) = 1 \qquad\qquad x \in \mathcal{R}$$

$$\mathcal{I}_i(x) \in \{0, 1\} \qquad\qquad i = 1,\dots,n;\ x \subset \mathcal{R}.$$

The authors conclude that the optimal boundary curves have the property that $\phi_i(x)/\phi_j(x)$ is constant. For the case where $\phi_i(x) = \|x - p_i\|^\alpha$ for any scalar $\alpha$, the Apollonius circle theorem [38] says that the boundary curves are circular arcs.

### 4.2.3 Further Uses of the Voronoi Paradigm

In addition to their complementary slackness properties, weighted Voronoi diagrams have proven useful in many other ways. For instance, [42] builds power Voronoi diagrams for mobile robots that partition a probability distribution $f$ equitably, as in [5]. However, rather than exploiting linear programming duality, their design a *decentralized* control scheme that is globally convergent and depends only on single-dimensional integrals. Convergence is established by a topological argument similar to Brouwer's fixed-point theorem.

The paper [29] uses multiplicatively weighted Voronoi diagrams for an application in logistics. They seek a multiplicatively weighted Voronoi diagram in which the cost function associated with each region is somewhat more complex because it involves both the mass of each district $\int_{D_i} f(x)\,dx$ as well as additional terms that approximate vehicular travel within each $D_i$. They introduce an intuitive fitting process that they prove is globally convergent.

The paper [44] uses multiplicative Voronoi diagrams for partitioning an abstract *information space*, as opposed to a physical region. Specifically, the objective is to find an effective representation of an information space, such as a set of documents, as a planar diagram that conveys relevant information. In this setting, each document is represented as a region in the plane, and there are two major objectives that should be considered in designing such a diagram effectively: first, documents containing similar content should be placed in close geographic proximity to one another. Second, documents with larger significance or relevance should be represented by regions that are larger than those corresponding to documents with less significance. The authors describe an iterative scheme that involves increasingly fine discretizations of the input region.

In an application to discrete choice models, [2] use power Voronoi diagrams in (possibly high) dimension to prove a remarkable bijection between two standard models of demand for differentiated products. Specifically, let $\lambda = (\lambda_1, \ldots, \lambda_n)$ denote the prices of $n$ variants of a differentiated product. An *aggregate demand system* (ADS) $\mathbb{F} : \mathbb{R}^n_+ \to \mathbb{R}^n_+$ is a vector-valued function $F(\lambda)$ such that $F_i(\lambda)$ represents the total demand for variant $i$ from a given population of consumers when the variants are priced according to $\lambda$.

An alternative model to the ADS is the *address model*, which is essentially a power Voronoi diagram in high dimension. More precisely, each of the $n$ product variants is represented as a point $p_i$ in a "characteristics space" $\mathbb{R}^m$, and there is a continuum of consumers distributed in $\mathbb{R}^m$ according to a density function $f$. Each consumer purchases one unit of the variant that offers the greatest utility, where the utility of a consumer located at $x \in \mathbb{R}^m$, purchasing variant $i$, is

$$u_i(x) = \alpha_i - \|x - p_i\|^2 - \lambda_i$$

where $\alpha_i$ is a perceived "quality index" of variant $i$ and $\lambda_i$ is the price of variant $i$ as in the ADS. Under the address model, we see that the total demand for variant $i$, written $\tilde{F}_i(\lambda)$, is

$$\tilde{F}_i(\lambda) = \iiint_{D_i} f(x) \, dx \,,$$

where $D_i$ is the power Voronoi cell of point $p_i$ with respect to the weight vector $\lambda - \alpha$.

The profound insight of [2] is that there exists an *equivalence* between the ADS and the address models demand, under certain natural conditions, such as gross substitution, constant aggregate demand, and invariance under scalar addition. Many standard discrete choice models satisfy these conditions, such as the logit, probit, linear probability, and CES models. Moreover, the equivalence established is *constructive*: the authors give a closed-form expression for the placement of points $p_i$ and the consumer density function $f$ in terms of $F$.

It turns out that Theorem 4.1 through Corollary 4.1 admit even further generality than that already described: we have already discussed generalizing the Euclidean distance function to more general cost functions $\phi$. In fact, [30] shows that essentially the same result holds for extremely general cost functions $\phi_i$, and that a proof can be obtained using a direct and simple minimization of a quadratic objective function.

## 4.3   The Ham Sandwich Paradigm

The *ham sandwich theorem* is a famous measure-theoretic result that says that, given $d$ measurable "objects" in $\mathbb{R}^d$, it is possible to divide all of them in half with a single $(d-1)$-dimensional hyperplane. The special case where $d = 2$ is called the *pancake theorem* and is easily proven by a bisection procedure involving a "rotating knife" [35]. The case $d = 3$ can also be proven intuitively [52], although the case for general $d$ requires the Borsuk–Ulam theorem [12]. Figure 4.6 shows three examples of *ham sandwich cuts* of different "objects" in the plane.

Both from the theoretical and algorithmic sides, the ham sandwich theorem admits many generalizations. One extension is the bisection of more than $d$ objects in $\mathbb{R}^d$ with more complex surfaces: given $\binom{k+d}{d} - 1$ measures in $\mathbb{R}^d$, there exists an algebraic surface of degree $k$ that bisects them all [49]. The case where $d = 2$ and both "objects" are sets of points (i.e., the case shown in 4.6b) is very well-studied, algorithmically speaking, as we will describe in the next section.

In most districting applications, one is interested in partitioning a territory into more than two pieces. Recursive application of the ham sandwich theorem guarantees that, when $n = 2^k$ and $k$ is an integer, it is always possible to partition $d$ measurable "objects" in $\mathbb{R}^d$ into $n$ convex pieces, such that each piece contains $1/n$ of the mass of all $d$ objects, as shown in Fig. 4.7. It is natural to ask if this statement holds for general $n$ (a similar statement was first conjectured in [33]), and

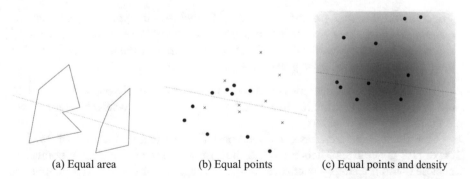

(a) Equal area                (b) Equal points                (c) Equal points and density

**Fig. 4.6** Ham sandwich cuts of three different pairs of "objects". In (**a**), we are given two shapes and the line cuts both simultaneously into pieces of equal area. In (**b**), the line cuts the two point sets simultaneously into halves. In (**c**), the line simultaneously bisects the point set and a probability density

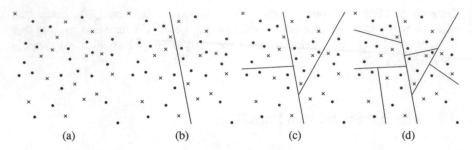

**Fig. 4.7** Given an initial collection of 16 "red" points and 24 "blue" points (**a**), we form a ham sandwich cut into two districts with 8 red and 12 blue (**b**). Each of these two pieces is further subdivided four districts containing 4 red and 6 blue (**c**), and finally, we subdivide those four into eight districts containing 2 red and 3 blue (**d**)

in fact the answer is affirmative, as proven in [11]. This influential result inspired a considerable amount of further study, as we summarize here.

### 4.3.1   Red and Blue Points

The paper [11] gives a fast algorithm for the following problem: given $gm$ red points and $gn$ blue points in the plane in general position, find a partition of the plane into $g$ convex districts, such that each district contains $m$ red points and $n$ blue points. Figure 4.7d shows the output for the case where $g = 8, m = 2, n = 3$. Although one can obtain such a partition by taking recursive ham sandwich cuts when $g$ is a power of 2 (precisely as is done in Fig. 4.7), it turns out that such an approach does not extend to general $g$, as is indicated in Fig. 4.8. Their major insight is the following theorem, which guarantees the existence of the desired partition for all $g$:

**Theorem 4.3** *For any $gm$ red points ($g \geq 2$) and $gn$ blue points in the plane in general position, there is a partition of the plane into 3 convex districts (one of which may be empty) such that the $i$th district contains $g_i m$ red points and $g_i n$ blue points, with $0 \leq g_1, g_2, g_3 < g$ and $g_1 + g_2 + g_3 = g$.*

The papers [32, 45] also give essentially the same result, although [11] gives an efficient algorithm (its running time is $\mathcal{O}(N^{4/3} \log^3 N \log g)$, where $N = g(m + n)$ is the total number of points).

Considerable research has been done on related variations and extensions. The paper [10] generalizes Theorem 4.3 to the case where the red and blue points are enclosed within a simply connected polygonal region and the goal is to obtain an equitable partition in which each district is *relatively convex* to the input region. A district $D_i$ of a polygonal region $\mathcal{R}$ is said to be *relatively convex* to $\mathcal{R}$ if, for every pair of points $x, y \in D_i$, the shortest path between $x$ and $y$ with respect to the geodesic induced by $\mathcal{R}$ is also contained in $D_i$; see Fig. 4.9. Another elegant

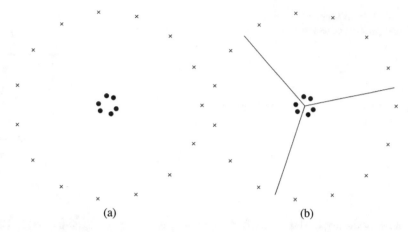

**Fig. 4.8** Suppose that we have $g = 3$ with $m = 5$ and $n = 2$. It is easy to see that we cannot obtain a partition into $g = 3$ pieces via recursive ham sandwich cuts because the blue points are concentrated in the center (**a**). However, it is possible to partition into three pieces, as shown in (**b**)

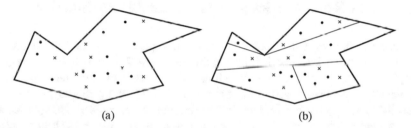

**Fig. 4.9** Given a simply connected polygonal region containing red and blue points (**a**), it is always to find a relatively convex partition (**b**) in the same sense as Theorem 4.3. Note that the uppermost district in (**b**) turns out to be the union of two triangles

variation studied in [9] addresses a rectilinear version: given any configuration of red and blue points as before, there exists an equitable partition consisting of at most $g - 1$ horizontal segments and $g - 1$ vertical segments, as shown in Fig. 4.10.

### 4.3.2 General Measures

Algorithms for partitioning more general measures than discrete and blue point sets have also been considered. The papers [1, 22] give efficient algorithms for the following problem: given a convex polygon $\mathcal{R}$ containing $n$ points, partition $\mathcal{R}$ into $n$ convex districts of equal area, such that each district contains one point. The proposed application is multi-depot vehicle routing; here each of the $n$ points represents the "depot," or start point, of a vehicle, and each vehicle is assigned to the

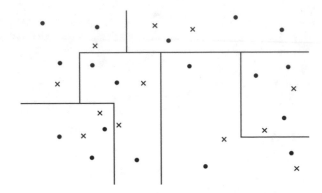

**Fig. 4.10** A rectilinear equitable partition of two point sets

district containing it. The case where $\mathcal{R}$ is simply connected is addressed in [20], where the goal is to obtain relatively convex districts as in [10].

## 4.4 Other Recursive Schemes for Equitable Partitioning

The basic principle of the ham sandwich paradigm is recursion: one obtains an equitable partition into districts by subdividing into 2 or 3 convex sub-regions, then further subdividing those sub-regions. Not surprisingly, such techniques arise frequently in geometric districting problems. For example, this is precisely the basis for *binary space partitioning* in 3d computer graphics [46]. One particularly elegant application arises in a recursive partitioning scheme in [41]: here we are given a collection of polygonal objects in the plane, and the goal is to design as few convex districts as possible such that each district contains only one object. When the polygonal objects are disjoint line segments, it turns out—somewhat surprisingly—that a simple randomized scheme results in a desired set of districts of size $\mathcal{O}(n \log n)$, where $n$ is the number of line segments.

The paper [3] describes simple approximation algorithms for the following problem: given a convex region $\mathcal{R}$ containing $n$ points, design districts $D_i$ of equal area so as to minimize the average distance between a point uniformly sampled in $\mathcal{R}$ and the facility that serves it. As an aside, the authors show how to divide $\mathcal{R}$ into convex districts of equal area while maximizing the "fatness" of the resulting regions within a constant factor. Their scheme consists of simply partitioning $\mathcal{R}$ recursively with either vertical or horizontal lines into proportions as close to one-half as possible.

Recently, Buot and Kano [19] have studied the *weighted* version of a family of problems closely related to ham sandwich partitioning: given sets of $p$ red and $q$ blue points in the plane with weights $\alpha > 0$ assigned to each red point and $\beta > 0$ to each blue point, the goal is to partition the plane into districts so that the total weight (from both sets) is equal. It turns out that the special case where $\alpha = 2$ and $\beta = 1$ is particularly relevant, and in particular, the authors show that for any configuration with total weight $2p + q = n\omega$ for some odd $\omega$, the plane can be subdivided into $n$ convex regions of weight $\omega$ if and only if $q \geq n$. They then give a recursive subdivision scheme for generating the desired partition.

The paper [28] interprets recursive partitioning in a creative way: given a pair of planar convex regions $A \subseteq B$ (a *pizza*), a partition is defined as a succession of double operations: a cut by a full straight line, followed by a Euclidean move of one of the resulting pieces; then the procedure is repeated. We seek a partition into $n$ districts so that each district contains the same amount of $A$ and of $B$. The authors prove that an equitable partition exists if and only if $n$ is even.

## 4.5   Cake Cutting

*Cake cutting*, also called *fair partitioning*, is an economic concept in which the objective is to divide a resource among a group of $n$ agents, each of whom has their own utility density function $f_i(\cdot)$ defined on the resource [16]. The goal is usually to divide the resource into $n$ pieces in such a way that each agent $i$ receives their "favorite" piece $R_j$, i.e., the piece that maximizes their utility, given by $\arg\max_j \int_{R_j} f_i(x)\,dx$. Such a division is said to be *envy-free*, in the sense that no agent desires any other agent's piece; see Fig. 4.11 for an example.

Cake cutting is the subject of an enormous volume of literature, although the underlying geometry of the resource being divided is not usually taken into account; typically one assumes that $\mathcal{R}$ is the unit interval and the goal is to obtain an envy-free subdivision that breaks $\mathcal{R}$ into as few pieces as possible. One exception is [31], which shows that it is possible to break a disputed territory into connected pieces among several other countries sharing a border with it in a proportional way. Recently, Segal-Halevi et al. [47] study the problem of cake cutting a rectangle or square. They demonstrate that even guaranteeing mere *proportionality*—requiring

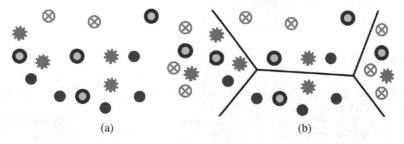

(a)                                      (b)

**Fig. 4.11**   An envy-free division of $\mathbb{R}^2$ among $n = 4$ agents. (**a**) We have 4 utility density functions $f_i$, represented here as atomic densities for visual clarity. In (**b**), the leftmost piece contains two units of one density (✳) and each of the other pieces contains only one unit of that density. Therefore, the agent corresponding to density ✳ prefers the left piece to any of the others. In this partition, it turns out that each agent's preferred piece can be uniquely assigned to that agent

that each agent receives at least $1/n$ of their total utility—is not guaranteed to exist when all pieces have to be squares or fat rectangles, although they also demonstrate that a constant-factor approximation to a proportional division is usually guaranteed to exist.

## 4.6 Equal Boundary Partitioning

A deceptively difficult problem originally posed on a blog in 2006 [37] is the following: given a convex planar shape $\mathcal{R}$ and an integer $n$, find a partition of $\mathcal{R}$ into $n$ convex pieces so that all pieces have equal area and equal perimeter, or determine that none exists. The problem as stated is still open, although it was proven in [14] that the statement is true when $n$ is a prime power via equivariant obstruction theory (which [54] likened to firing a cannon at a sparrow). Note that when we merely require that each piece has the same amount of the boundary of $\mathcal{R}$, we are guaranteed an affirmative result by the ham sandwich theorem. The paper [34] extends this analysis to convex shapes in $\mathbb{R}^d$.

## 4.7 Conclusions

As we have seen, for many practical districting problems, it may be desirable to formulate the problems geometrically and then use a computational geometric method to solve it, as opposed to a discrete combinatorial approach. Formulating an instance of such a problem requires an intrinsically interdisciplinary approach, combining elements from computational geometry, geometric probability theory, and optimization. Indeed, despite the obvious connection between geometry and operations research, it seems that many simple and fundamental problems remain that have yet to be fully understood.

## References

1. Adjiashvili, D., Peleg, D.: Equal-area locus-based convex polygon decomposition. Theor. Comput. Sci. **411**(14–15), 1648–1667 (2010)
2. Anderson, S.P., De Palma, A., Thisse, J.F.: Demand for differentiated products, discrete choice models, and the characteristics approach. Rev. Econ. Stud. **56**(1), 21–35 (1989)
3. Aronov, B., Carmi, P., Katz, M.J.: Minimum-cost load-balancing partitions. Algorithmica **54**(3), 318–336 (2009)
4. Aurenhammer, F.: Voronoi diagrams—a survey of a fundamental geometric data structure. ACM Comput. Surv. **23**(3), 345–405 (1991)
5. Aurenhammer, F., Hoffmann, F., Aronov, B.: Minkowski-type theorems and least-squares clustering. Algorithmica **20**(1), 61–76 (1998)

6. Barber, C.B., Dobkin, D.P., Huhdanpaa, H.: The quickhull algorithm for convex hulls. ACM Trans. Math. Softw. **22**(4), 469–483 (1996)
7. Basu, A., Mitchell, J.S., Sabhnani, G.K.: Geometric algorithms for optimal airspace design and air traffic controller workload balancing. J. Exp. Algorithmics **14**, 3–31 (2009)
8. Beardwood, J., Halton, J.H., Hammersley, J.M.: The shortest path through many points. Math. Proc. Camb. Philos. Soc. **55**(4), 299–327 (1959)
9. Bereg, S.: Orthogonal equipartitions. Comput. Geom. **42**(4), 305–314 (2009)
10. Bereg, S., Bose, P., Kirkpatrick, D.: Equitable subdivisions within polygonal regions. Comput. Geom. **34**(1), 20–27 (2006)
11. Bespamyatnikh, S., Kirkpatrick, D., Snoeyink, J.: Generalizing ham sandwich cuts to equitable subdivisions. Discret. Comput. Geom. **24**(4), 605–622 (2000)
12. Beyer, W.A., Zardecki, A.: The early history of the ham sandwich theorem. Am. Math. Mon. **111**(1), 58–61 (2004)
13. Bichot, C.E., Siarry, P. (eds.) Graph Partitioning. Wiley, New York (2011)
14. Blagojević, P.V., Ziegler, G.M.: Convex equipartitions via equivariant obstruction theory. Isr. J. Math. **200**(1), 49–77 (2014)
15. Boyd, T.D., Jameson, M.H.: Urban and rural land division in ancient Greece. Hesperia J. Am. Sch. Class. Stud. Athens **50**(4), 327–342 (1981)
16. Brams, S.J., Taylor, A.D.: Fair Division: From Cake-Cutting to Dispute Resolution. Cambridge University, Cambridge (1996)
17. Brieden, A., Gritzmann, P.: On optimal weighted balanced clusterings: gravity bodies and power diagrams. SIAM J. Discret. Math. **26**(2), 415–434 (2012)
18. Brieden, A., Gritzmann, P., Klemm, F.: Constrained clustering via diagrams: a unified theory and its application to electoral district design. Eur. J. Oper. Res. **263**(1), 18–34 (2017)
19. Buot, J., Kano, M.: Weight-equitable subdivision of red and blue points in the plane. Int. J. Comput. Geom. Appl. **28**(1), 39–56 (2018)
20. Carlsson, J.G.: Dividing a territory among several vehicles. INFORMS J. Comput. **24**(4), 565–577 (2012)
21. Carlsson, J.G., Devulapalli, R.: Dividing a territory among several facilities. INFORMS J. Comput. **25**(4), 730–742 (2012)
22. Carlsson, J.G., Armbruster, B., Ye, Y.: Finding equitable convex partitions of points in a polygon efficiently. ACM Trans. Algorithms **6**(4), 72 (2010)
23. Carlsson, J.G., Carlsson, E., Devulapalli, R.: Shadow prices in territory division. Netw. Spat. Econ. **16**(3), 893–931 (2016)
24. Cortés, J.: Coverage optimization and spatial load balancing by robotic sensor networks. IEEE Trans. Autom. Control **55**(3), 749–754 (2010)
25. Didandeh, A., Bigham, B.S., Khosravian, M., Moghaddam, F.B.: Using Voronoi diagrams to solve a hybrid facility location problem with attentive facilities. Inf. Sci. **234**(1), 203–216 (2013)
26. Drezner, T., Drezner, Z.: Voronoi diagrams with overlapping regions. OR Spectr. **35**(3), 543–561 (2013)
27. Fortune, S.: A sweepline algorithm for Voronoi diagrams. Algorithmica **2**(1–4), 153 (1987)
28. Fruchard, A., Magazinov, A.: Fair partitioning by straight lines. In: Adiprasito, K., Bárány, I., Vilcu, C. (eds.) Convexity and Discrete Geometry Including Graph Theory. Springer Proceedings in Mathematics and Statistics, vol. 148, pp. 161–165. Springer, Cham (2016)
29. Galvão, L.C., Novaes, A.G., De Cursi, J.S., Souza, J.C.: A multiplicatively-weighted Voronoi diagram approach to logistics districting. Comput. Oper. Res. **33**(1), 93–114 (2006)
30. Geiß, D., Klein, R., Penninger, R., Rote, G.: Reprint of: optimally solving a transportation problem using Voronoi diagrams. Comput. Geom. **47**(3), 499–506 (2014)
31. Hill, T.P.: Determining a fair border. Am. Math. Mon. **90**(7), 438–442 (1983)
32. Ito, H., Uehara, H., Yokoyama, M.: 2-dimension ham sandwich theorem for partitioning into three convex pieces. In: Akiyama, J., Kano, M., Urabe, M. (eds.) Discrete and Computational Geometry: Japanese Conference on Discrete and Computational Geometry (JCDCG 1998). Lecture Notes in Computer Science, vol. 1763, pp. 129–157. Springer, Berlin (2000)

33. Kaneko, A., Kano, M.: Balanced partitions of two sets of points in the plane. Comput. Geom. **13**(4), 253–261 (1999)
34. Karasev, R., Hubard, A., Aronov, B.: Convex equipartitions: the spicy chicken theorem. Geom. Dedicata **170**(1), 263–279 (2014)
35. Kinsey, L.C.: Topology of Surfaces. Springer, New York (1993)
36. Marx, D., Pilipczuk, M.: Optimal parameterized algorithms for planar facility location problems using Voronoi diagrams. In: Bansal, N., Finocchi, I. (eds.) Algorithms—ESA 2015. Lecture Notes in Computer Science, vol. 9294, pp. 865–877. Springer, Berlin (2015)
37. Nandakumar, R.: Fair Partitions (2018). http://nandacumar.blogspot.com/2006/09/cutting-shapes.html. Accessed 1 July 2018
38. Ogilvy, C.S.: Excursions in Geometry. Dover, Mineola (1990)
39. Okabe, A., Suzuki, A.: Locational optimization problems solved through Voronoi diagrams. Eur. J. Oper. Res. **98**(3), 445–456 (1997)
40. Okabe, A., Boots, B., Sugihara, K., Chiu, S.N.: Spatial Tessellations: Concepts and Applications of Voronoi Diagrams, 2nd edn. Wiley, Chichester (2000)
41. Paterson, M.S., Yao, F.F.: Efficient binary space partitions for hidden-surface removal and solid modeling. Discret. Comput. Geom. **5**(5), 485–503 (1990)
42. Pavone, M., Arsie, A., Frazzoli, E., Bullo, F.: Distributed algorithms for environment partitioning in mobile robotic networks. IEEE Trans. Autom. Control **56**(8), 1834–1848 (2011)
43. Pavone, M., Frazzoli, E., Bullo, F.: Adaptive and distributed algorithms for vehicle routing in a stochastic and dynamic environment. IEEE Trans. Autom. Control **56**(6), 1259–1274 (2011)
44. Reitsma, R., Trubin, S., Mortensen, E.: Weight-proportional space partitioning using adaptive Voronoi diagrams. Geoinformatica **11**(3), 383–405 (2007)
45. Sakai, T.: Radial partitions of point sets in $R^2$. In: Proceedings of the Japan Conference on Discrete and Computational Geometry '98 (Collection of Extended Abstracts), pp. 74–78. Tokai University, Tokyo (1998)
46. Schumacker, R.A., Brand, B., Gilliland, M.G., Sharp, W.H.: Study for applying computer-generated images to visual simulation. Technical report AFHRL-TR-69-14, US Air Force Human Resources Lab, Brooks Air Force Base, USA (1969)
47. Segal-Halevi, E., Nitzan, S., Hassidim, A., Aumann, Y.: Fair and square: cake-cutting in two dimensions. J. Math. Econ. **70**(1), 1–28 (2017)
48. Shamos, M.I., Hoey, D.: Closest-point problems. In: Proceedings of the 16th Annual Symposium on Foundations of Computer Science, pp. 151–162. IEEE, Berkeley (1975)
49. Smith, W.D., Wormald, N.C.: Geometric separator theorems and applications. In: Proceedings of the 39th Annual Symposium on Foundations of Computer Science, pp. 232–243. IEEE Computer Society, Washington (1998)
50. Soberón, P.: Gerrymandering, sandwiches, and topology. Not. AMS **64**(9), 1010–1013 (2017)
51. Svec, L., Burden, S., Dilley, A.: Applying Voronoi diagrams to the redistricting problem. UMAP J. **28**(3), 313–329 (2007)
52. Weisstein, E.W.: Ham sandwich theorem. In: MathWorld—A Wolfram Web Resource (2018). http://mathworld.wolfram.com/HamSandwichTheorem.html. Accessed 1 July 2018
53. Xue, M.: Airspace sector redesign based on Voronoi diagrams. J. Aerosp. Comput. Inf. Commun. **6**(12), 624–634 (2009)
54. Ziegler, G.M.: Cannons at sparrows. Eur. Math. Soc. Newsl. **95**, 25–31 (2015)

# Part II
# Theory, Models, and Algorithms

# Chapter 5
# Bounding Procedures and Exact Solutions for a Class of Territory Design Problems

Juan A. Díaz, Dolores E. Luna, and María G. Sandoval

**Abstract** In this chapter, we present and evaluate exact methods and lower and upper bounding procedures for a class of territory design problems. Most territory design problems, as the one studied in this chapter, consider requirements of compactness, contiguity, and balance with respect to one or more activity measures, for example, number of customers and sales volume in the case of commercial territories, voting potential equality in the case of political territories, workload balance when designing service territories, etc. To obtain solutions with compact territories, a minisum objective function equivalent to the objective function of the $p$-median problem is used. The exact solution methods presented here use different relaxations of integer linear programming formulations of the problem. Additionally, two methodologies to obtain upper bounds (feasible solutions) are presented. The first one uses the relaxation of an integer quadratic programming formulation. The second methodology obtains feasible solutions using a primal heuristic within the framework of a subgradient optimization algorithm to solve a Lagrangian dual that also provides lower bounds for the optimal solution. Instances obtained from the literature are used to evaluate and compare the different methodologies presented.

## 5.1 A Class of Territory Design Problems

Territory design problems consist of grouping small geographic areas, which are known as basic units, into groups that are normally called districts in such a way that they comply with a set of planning criteria. Among the most common planning criteria, it is desired to have districts that are compact, contiguous, and balanced with respect to one or several activities [4]. It is said that a district is compact if

J. A. Díaz (✉) · D. E. Luna · M. G. Sandoval
Universidad de las Américas Puebla, Ex Hacienda Santa Catarina Mártir, Cholula, Puebla, Mexico
e-mail: juana.diaz@udlap.mx; dolorese.luna@udlap.mx; maria.sandovalel@udlap.mx

© Springer Nature Switzerland AG 2020
R. Z. Ríos-Mercado (ed.), *Optimal Districting and Territory Design*, International Series in Operations Research & Management Science 284,
https://doi.org/10.1007/978-3-030-34312-5_5

its shape is roughly round, does not have holes nor is distorted. It is considered that a district is contiguous if it is possible to travel between all the basic units that compose it without having to leave the district. Finally, it is said that the districts are balanced with respect to a certain activity measure, if their sizes are similar. The first two characteristics, compactness and contiguity, help reduce the travel times of the people who serve each district.

This type of problem can be applied in practice in many situations. Some examples are the design of electoral districts [11, 15], school districts [3, 8], sales and commercial territories [18], police districts [4], electrical power districts [1], territories to offer garbage collection services or snow removal in winter [14, 20], etc. In some applications, such as in political districting, the criterion of contiguity is mandatory. Examples of balance criteria, when designing commercial territories, are demand volume, number of costumers, workload, etc. In the case of political districting, balance criteria are population equality, satisfy the neutrality condition, etc. [13]

Different approaches have been proposed in the literature to model and to find upper and lower bounds for the territory design problem. Also, different mathematical expressions have been used to model the compactness, contiguity, and balance characteristics. Kalcsics [12] gives a broad overview of typical criteria and restrictions used in several districting studies and presents different approaches to quantify and model these criteria. Surveys can be found in both Kalcsics et al. [13] and in Ricca et al. [16]. D'Amico et al. [4] study a police districting problem where the objectives are the effective use of patrol cars and the workload balance between officers in different districts. They model this problem as a constrained graph-partitioning problem and propose a simulated annealing heuristic to provide feasible solutions to the problem. Bozkaya et al. [2] study the political districting problem. They propose a model that assigns weights to a set of criteria and a tabu search and adaptive memory heuristic to find feasible solutions for the studied problem. Among the criteria used in their model are: population equality, compactness, and socio-economic homogeneity. Fernández et al. [9] study a districting problem motivated by the recycling directive WEEE (waste electrical and electronic equipment) of the European Union. The problem studied in this research is different from classical territory design because the territories should be as geographically dispersed as possible. They identify an appropriate measure for the dispersion of a territory. Then, they present a mathematical programming model for the problem and a solution method based on the GRASP methodology. Their computational results exemplify the suitability of the model and assess the effectiveness of the heuristic. Ríos-Mercado and Fernández [17] model the problem as a vertex $p$-center problem with multiple capacity constraints and propose a reactive GRASP algorithm to obtain feasible solutions for the problem. The purpose of model the objective function as a $p$-center problem is to measure territory dispersion. Ríos-Mercado and López-Pérez [18] present a heuristic based on the iterative solution of a relaxed mixed integer linear programming problem. Elizondo-Amaya et al. [7] model the problem using a $p$-center objective function and develop a dual bounding scheme for a commercial territory design problem considering balance and compactness requirements. Their algorithm performs a binary search over a set of radii extracted from the distances matrix and solves each of them using a Lagrangian dual problem based on a maximal demand covering problem. Hasse and Müller [10] study the sales force deployment problem that involves the concurrent resolution of four interrelated subproblems: sizing of the sales force, sales representatives' locations, sales territory alignment, and sales resource allocation. Their objective is to maximize the total profit. They use a concave sales response function and propose a model formulation with an infinite number of binary variables. The linear relaxation of the model is solved using column generation. For the optimal objective function value of the linear relaxation, an upper bound is provided and to obtain a tight gap for the objective function value of the optimal integer solution they use a Branch-and-Price approach.

Salazar-Aguilar et al. [19] study a territory design problem with balance constraints associated with two different activity measures (number of customers and sales volume) and connectivity constraints. They consider two different objective functions to model territory compactness: a minisum objective and a minimax objective. When the minisum objective is used, the problem can be viewed as a $p$-median problem with balance and connectivity constraints. On the other hand, when the minimax objective is used, the problem can be viewed as a $p$-center problem with balance and connectivity constraints. They propose several formulations for the studied problems. These formulations include linear integer programming models and also quadratic integer programming formulations that involve a smaller number of variables than the linear ones. They also introduce an exact solution framework for this problem, that is based on branch and bound and a cut generation strategy. Their method is empirically evaluated using several instances.

Díaz and Luna [5] study a relaxation of the territory design problem with the minisum objective studied in [19], where connectivity constraints are dropped. Therefore, the solution of the relaxation studied in that work provides good feasible balanced clusters that can be useful as a starting point for heuristic methods for the territory design problem with connectivity constraints. They include balance constraints for two activities as measures of equality as those presented in different situations. These balance constraints are used to model different equity measures commonly presented in territory design problems. Some examples are population equality and impartiality for political districting problems, workload balance for service districting problems, revenue balance for electrical power districting problems, number of customers assigned to commercial districts, etc. A Lagrangian relaxation scheme is proposed to obtain lower bounds and a primal heuristic is proposed to obtain upper bounds for the problem. The heuristic procedure first provides feasible allocations given a set of medians, and then improves the solutions using a median exchange procedure. In [7] a similar relaxation with the minimax problem is presented.

In this chapter, we present methods to obtain lower and upper bounds for a territory design problem with the following characteristics. A minisum objective function based on the $p$-median dispersion measure is used to ensure compactness in the solution. This solution must also satisfy both, connectivity constraints and balance constraints with respect to two activity measures. First, the procedures for obtaining feasible bounds or optimal solutions proposed in [19] are presented. These procedures use relaxations of an integer linear programming model and an integer quadratic programming model for the problem studied here. The Lagrangian relaxation proposed in [5] to obtain bounds for the problem is also presented. Then, we propose a methodology to obtain feasible bounds or optimal solutions for the problem. We assess the quality of the solutions obtained by all methods using a set of problem instances from the literature. According to the computational experience, the methods proposed in this chapter provide very good quality solutions in very competitive CPU times.

The rest of the chapter is organized as follows: Sect. 5.2 describes the problem and presents the models for the problem studied in this chapter. Section 5.3 describes procedures to find upper and lower bounds as well as exact solution methods for the problem. Section 5.4 presents the computational experience to evaluate the performance of the different methods. Finally, Sect. 5.5 provides some concluding remarks.

## 5.2   Problem Description and Mathematical Models

In this section, we present different formulations from the literature to represent the class of territory design problems studied in this work. We now describe two formulations for the territory design problem with compactness, connectivity, and balance constraints studied in this chapter. These models are proposed in [19]. Let $G = (V, E)$ be a graph with $V = \{1, \ldots, n\}$ a set of basic units (BUs) and $E$ a set of edges representing adjacency between BUs. Also, let $p$ be the number of territories in which we wish to divide the set of BUs, and it is required that each BU is assigned to only one territory. For each edge $\{i, j\} \in E$, $d_{ij}$ denotes the distance between basic units $i$ and $j$. Also, let $A$ be the set of activity measures, such as population, demand, area, etc. For each $j \in V$ and $a \in A$, $w_j^a$, denotes the value associated to activity $a$ in node $j$. As it is impossible to have perfectly balanced territories with respect to each activity $a \in A$, to model balance, a tolerance parameter $\tau^a$ for activity $a$ is used. This parameter measures the relative deviation from the average territory size with respect to activity $a \in A$. The average for activity $a$ is given by

$$\mu^a = \sum_{j \in V} w_j^a / p.$$

The single allocation of each BU to one of the $p$ medians selected is associated with a partition of the set of nodes $V$ into $p$ balanced clusters with respect to each activity measure. Therefore, an allocation of nodes to the selected median nodes will be feasible if, for each activity $a \in A$, the size of each cluster is between $(1 - \tau^a)\mu^a$

and $(1 + \tau^a)\mu^a$. It is also required that for each $i \in V$ and $j \in V$ assigned to the same territory, there exists a path between them totally contained in the territory in order to achieve connectivity. Finally, to measure territory compactness a minisum objective function based on the $p$-median problem is used. Therefore, the problem studied in this chapter is to find a partition of $V$ in $p$ territories according to planning requirements of balancing and connectivity that minimizes the total sum of distances between the medians and the BUs allocated to them.

## 5.2.1 *Integer Linear Programming Model*

Now, we describe the integer linear model for the problem introduced in [19]. Let us define $N^i$ as the set of nodes adjacent to node $i$, that is:

$$N^i = \{j \in V : \{i, j\} \in E\}, i \in V$$

and let us define the following decision variables:

$$x_{ij} = \begin{cases} 1, & \text{if basic unit } j \text{ is allocated to median } i, \\ 0, & \text{otherwise.} \end{cases}$$

for all $i, j \in V$.

Therefore, the mixed integer linear model can be formulated as:

$$(\text{TDP}_{\text{IP}}) \quad \min \quad \sum_{i \in V} \sum_{j \in V} d_{ij} x_{ij} \tag{5.1a}$$

$$\text{subject to} \quad \sum_{i \in V} x_{ii} = p \tag{5.1b}$$

$$\sum_{i \in V} x_{ij} = 1 \qquad\qquad j \in V \tag{5.1c}$$

$$\sum_{j \in V} w_j^a x_{ij} \geq (1 - \tau^a)\mu^a x_{ii} \qquad i \in V, a \in A \tag{5.1d}$$

$$\sum_{j \in V} w_j^a x_{ij} \leq (1 + \tau^a)\mu^a x_{ii} \qquad i \in V, a \in A \tag{5.1e}$$

$$\sum_{j \in \bigcup_{v \in S}(N^v \setminus S)} x_{ij} - \sum_{j \in S} x_{ij} \geq 1 - |S| \quad i \in V, S \subset \left(V \setminus \left(N^i \cup \{i\}\right)\right) \tag{5.1f}$$

$$x_{ij} \in \{0, 1\} \qquad\qquad i, j \in V \tag{5.1g}$$

The objective function (5.1a) minimizes the sum of the distances between each BU and the location of the median it is allocated to. This objective function tries to minimize dispersion which is equivalent to maximizing compactness. Constraint (5.1b) assures the creation of exactly $p$ territories. Constraints (5.1c) ensure that each node is assigned to only one territory. Constraints (5.1d) and (5.1e) represent territory balance with respect to each activity measure as they establish that the size of each territory must lie within the specified lower and upper bound of that activity (measured by the tolerance parameter $\tau^a$) around its average size $\mu^a$. These also assure that no BU is assigned to node $i$ if $i$ is not a median. Constraints (5.1f) guarantee territory connectivity. They ensure that for any given subset $S$ of BUs assigned to center $i$, not containing BU $i$, there must be an arc between $S$ and the set containing $i$. These constraints are similar to the subtour elimination constraints in the traveling salesman problem. It is important to note that there is an exponential number of such constraints so they cannot be explicitly written out.

## 5.2.2   Lagrangian Relaxation

Díaz and Luna [5] study the vertex $p$-median problem with balance constraints, that is a relaxation of the TDP$_{IP}$ model previously presented, where the connectivity constraints have been dropped. To obtain lower bounds for that problem they relax the assignment constraints (5.1c) in a Lagrangian fashion as follows: Let $\lambda \in \mathbb{R}^{|V|}$ be the vector of Lagrangian multipliers associated with the assignment constraints (5.1c). Then, the Lagrangian relaxation can be stated as:

$$(LR(\lambda)) \qquad \text{minimize} \quad z_{LR}(\lambda) = \sum_{i \in V} \sum_{j \in V} d_{ij} x_{ij} + \sum_{j \in V} \lambda_j \left( 1 - \sum_{i \in V} x_{ij} \right)$$

$$\text{subject to} \quad (5.1b), (5.1d), (5.1e) \text{ and } (5.1g)$$

By dualizing the assignment constraints in a Lagrangian fashion, the problem decomposes into $|V|$ subproblems, one for each potential median node $i \in V$. In [5] a procedure is proposed to compute the optimal solution for $LR(\lambda)$. Then, they use a subgradient optimization algorithm to solve the Lagrangian dual to find the best dual bound for the problem. Also, within the subgradient procedure a primal heuristic is used to find upper bounds for the optimal solution of the problem.

### 5.2.3 Integer Quadratic Programming Model

The integer quadratic programming model introduced in [19] reduces the number of binary variables from $n^2$ to $2np$. Let $Q = \{1, 2, \ldots, p\}$ be the set of territory indices and let:

$$z_{jq} = \begin{cases} 1, & \text{if basic unit } j \text{ is allocated to territory } q, \\ 0, & \text{otherwise.} \end{cases}$$

for all $j \in V, q \in Q$, and

$$y_{iq} = \begin{cases} 1, & \text{if basic unit } i \text{ is the median of territory } q, \\ 0, & \text{otherwise.} \end{cases}$$

for all $i \in V, q \in Q$.

Then, the equivalence between the variables in the linear model and the variables in the quadratic model is given by:

$$x_{ij} = \sum_{q \in Q} z_{jq} y_{iq}$$

Therefore, the integer quadratic programming model can be formulated as:

$$(\text{TDP}_{\text{QIP}}) \quad \min \quad \sum_{q \in Q} \sum_{i, j \in V} d_{ij} z_{jq} y_{iq} \tag{5.2a}$$

$$\text{subject to} \quad \sum_{i \in V} y_{iq} = 1 \qquad q \in Q \tag{5.2b}$$

$$\sum_{q \in Q} z_{jq} = 1 \qquad j \in V \tag{5.2c}$$

$$z_{jq} \geq y_{jq} \qquad q \in Q, j \in V \tag{5.2d}$$

$$\sum_{j \in V} w_j^a z_{jq} \geq (1 - \tau^a) \mu^a \qquad q \in Q, a \in A \tag{5.2e}$$

$$\sum_{j \in V} w_j^a z_{jq} \leq (1 + \tau^a) \mu^a \qquad q \in Q, a \in A \tag{5.2f}$$

$$\sum_{q \in Q} \sum_{j \in \bigcup_{v \in S}(N^v \setminus S)} z_{jq} y_{iq}$$

$$- \sum_{q \in Q} \sum_{j \in S} z_{jq} y_{iq} \geq 1 - |S| \quad i \in V, S \subset \left( V \setminus \left( N^i \cup \{i\} \right) \right) \tag{5.2g}$$

$$z_{jq} \in \{0, 1\} \qquad\qquad q \in Q,\, j \in V \qquad (5.2h)$$

$$y_{iq} \in \{0, 1\} \qquad\qquad q \in Q,\, i \in V \qquad (5.2i)$$

The TDP$_{QIP}$ (territory design problem based on quadratic integer programming) model uses an equivalent dispersion measure to that of TDP$_{IP}$ model (see (5.2a)). Constraints (5.2b) guarantee that there is only one median for each territory. Constraints (5.2c) assure that each BU is assigned to only one territory. Constraints (5.2d) establish that BU $j$ cannot be the median of $q$ if $j$ is not assigned to $q$. According to Proposition 2 in Domínguez and Muñoz [6], constraints (5.2b) and (5.2c) guarantee the assignment, and therefore constraints (5.2d) are not needed. However, as in [19], these are shown for model completeness. The set of constraints (5.2e) and (5.2f) assure territory balance, in a similar way as in the TDP$_{IP}$ model. The quadratic constraints (5.2g) assure connectivity. Again, there is an exponential number of these constraints.

## 5.3   Solution Methods and Bounding Procedures

In this section, we describe the algorithmic approaches to solve or to find upper and lower bounds for the territory design problem. All solution methods described consider relaxed formulations with respect to the connectivity constraints. The difficulty to solve the formulations proposed in the previous section lies in the existence of an exponential number of these constraints. When these constraints are relaxed, clusters of nodes that satisfy the rest of the constraints (single assignment, balance with respect to activity measures, etc.) are obtained from the optimal solutions of the relaxed problem. To find out if these solutions also satisfy the connectivity constraints, a separation problem can be solved to verify if the graph induced by the set of nodes of each cluster contains a single connected component. When it does, the optimal solution of the relaxed problem is also optimal for the TDP. Otherwise, a cut generation procedure can be used to incorporate violated connectivity constraints to the relaxed formulations associated with connected components whose node set does not contain the median of their cluster. All connected components of the graph induced by a cluster of nodes can be easily obtained with a deep-first search (DFS) procedure. Let $G = (V, E)$ be an undirected graph. The pseudocodes depicted in Algorithms 1 and 2 detail how all connected components of a graph can be found. Algorithm 1 describes a function to obtain all connected components of a graph and Algorithm 2 shows a recursive procedure to traverse all nodes of a connected component.

We first describe the algorithmic approaches proposed in [19] where the linear integer programming formulation TDP$_{IP}$ and the quadratic integer programming formulation TDP$_{QIP}$ are used. Since both formulations contain an exponential

---

**Algorithm 1** Procedure to find all connected components of a graph

---

**function** FINDCONNECTEDCOMPONENTS($G = (V, E)$)
    Let $Visited_v$, $v \in V$, be a boolean value which indicates if a node has been traversed,
    and $CC$ the set of nodes subsets associated to connected components of $G$
    **for all** ($v \in V$) **do**
        $Visited_v := $ **false**
    **end for**
    $CC := \emptyset$
    **for all** ($v \in V$) **do**
        **if** (**not** $Visited_v$) **then**
            $S := \emptyset$
            $DFS(G, v, Visited, S)$
        **end if**
        $CC := CC \cup \{S\}$
    **end for**
    **return** $CC$
**end function**

---

**Algorithm 2** Recursive procedure to traverse all nodes of a connected component

---

**procedure** $DFS(G, v, Visited, S)$
    $Visited_v := $ **true**
    $S := S \cup \{v\}$
    **for all** ($u \in V : \{u, v\} \in E$) **do**
        **if** (**not** $Visited_u$) **then**
            $DFS(G, u, Visited, S)$
        **end if**
    **end for**
**end procedure**

---

number of connectivity constraints, for each formulation of the problem, a relaxed formulation is used where the connectivity constraints are relaxed, and these formulations are coupled with a cut generation procedure to iteratively identify and add violated connectivity constraints in the optimal solution of the relaxed problem, until an optimal solution (or a local optimal solution if the quadratic programming formulation is used) is obtained.

In the case of the linear integer programming formulation TDP$_{IP}$, the authors consider a relaxed formulation that only includes the connectivity constraints associated to subsets of BUs of size one (see constraints (5.3)). Authors comment, that according to empirical tests, a very large proportion of unconnected optimal solutions of the relaxed model, are associated to subsets of cardinality one.

$$(TDP_{IPR_1}) \qquad \min \quad \sum_{i,j \in V} d_{ij} x_{ij}$$

$$\text{subject to} \quad \sum_{i \in V} x_{ii} = p$$

$$\sum_{i \in V} x_{ij} = 1 \qquad\qquad j \in V$$

$$\sum_{j \in V} w_j^a x_{ij} \geq (1 - \tau^a)\mu^a x_{ii} \qquad i \in V, a \in A$$

$$\sum_{j \in V} w_j^a x_{ij} \leq (1 + \tau^a)\mu^a x_{ii} \qquad i \in V, a \in A$$

$$\sum_{l \in N^j} x_{il} \geq x_{ij} \qquad\qquad i \in V, j \in V \setminus \left(\{i\} \cup N^i\right)$$

$$\tag{5.3}$$

$$x_{ij} \in \{0, 1\} \qquad\qquad i, j \in V$$

In [19] they also consider a relaxation of the quadratic integer program formulation $\text{TDP}_{\text{QIP}}$ where the connectivity constraints are relaxed.

If in $\text{TDP}_{\text{QIP}}$, we relax the set of constraints (5.2g), that guarantee the connectivity of the solution, the model can be solved to obtain a lower bound for the model $\text{TDP}_{\text{QIP}}$. The relaxed model can be reinforced by adding restrictions that do not allow territories of size 1. That is:

$$\sum_{i \in N^j} z_{iq} \geq z_{jq} \quad q \in Q, j \in V$$

This condition is satisfied if and only if $w_j^a < (1 - \tau^a)\mu^a$ for each $j \in V, a \in A$, and is true in this particular case because the data instances used to test the methods in this chapter always satisfy it. Then, the relaxed model can be formulated as:

$$(\text{TDP}_{\text{QIPR}}) \qquad \min \quad \sum_{q \in Q} \sum_{i,j \in V} d_{ij} z_{jq} y_{iq}$$

$$\text{subject to} \quad \sum_{i \in V} y_{iq} = 1 \qquad\qquad q \in Q$$

$$\sum_{q \in Q} z_{jq} = 1 \qquad\qquad j \in V$$

$$z_{jq} \geq y_{jq} \qquad\qquad q \in Q, j \in V$$

$$\sum_{j \in V} w_j^a z_{jq} \geq (1 - \tau^a)\mu^a \qquad q \in Q, a \in A$$

$$\sum_{j \in V} w_j^a z_{jq} \leq (1 + \tau^a)\mu^a \qquad q \in Q, a \in A$$

$$\sum_{i \in N^j} z_{iq} \geq z_{jq} \qquad\qquad q \in Q, \, j \in V \qquad (5.4)$$

$$z_{jq} \in \{0, 1\} \qquad\qquad q \in Q, \, j \in V$$

$$y_{iq} \in \{0, 1\} \qquad\qquad q \in Q, \, i \in V$$

The solution method proposed in [19] can use both of the relaxations of the problem mentioned above. It is an iterative procedure. At each iteration, one of the relaxed problems is solved and the set of $p$ clusters (territories) associated to the optimal solution of the relaxed problem is obtained. Then, all connected components, in the graphs induced by the set of nodes of each cluster, are identified. Let $C_1, C_2, \ldots, C_p$ be the set of the $p$ node clusters obtained in the optimal solution of the relaxed formulation, where $m_1, m_2, \ldots, m_p$ are the corresponding medians of each of the $p$ node clusters (that is, for each $k = 1, \ldots, p, x_{m_k,j} = 1$ for each $j \in C_k$.) For each $k, k = 1, \ldots, p$, we obtain the connected components of its respective induced graph, $G_k = (C_k, E(C_k))$, where $E(C_k) = \{\{i, j\} \in E : i, j \in C_k\}$ and $E = \bigcup_{j \in V} \{\{j, k\} : k \in N^j\}$. Let $S_1, \ldots, S_t$ be the collection of node subsets associated to the connected components identified in the induced graphs of the node clusters. If node subset $S_u$ does not contain the median of its cluster (that is, for all $j \in S_u, x_{rj} = 1$ but $r \notin S_u$, where $r$ is the median of the nodes in $S_u$), then $S_u$ is a node subset associated to a violated connectivity constraint, and it is added to the set $S^*$, where $S^*$ contains all node subsets associated to violated connectivity constraints in the optimal solution of the relaxation considered. If $S^* = \emptyset$ the iterative procedure terminates, and the optimal solution of the relaxed formulation is also optimal for the territory design problem. Otherwise, the connectivity constraints corresponding to each node subset in $S^*$ are added to the relaxed formulation and the iterative procedure is restarted. This procedure is depicted in Algorithm 3.

We propose a procedure, that is, similar to the one proposed in [19], to solve the territory design problem that uses the integer programming formulation $\text{TPD}_{\text{IP}_2}$. It is an iterative approach. At each iteration, a relaxed formulation is used to find a feasible solution with respect to constraints (5.1b)–(5.1e), (5.1g), and (5.5),

$$(\text{TDP}_{\text{IPR}_2}) \qquad \min \quad \sum_{i,j \in V} d_{ij} x_{ij}$$

$$\text{subject to} \quad \sum_{i \in V} x_{ii} = p$$

$$\sum_{i \in V} x_{ij} = 1 \qquad\qquad j \in V$$

$$\sum_{j \in V} w_j^a x_{ij} \geq (1 - \tau^a) \mu^a x_{ii} \qquad i \in V, \, a \in A$$

$$\sum_{j \in V} w_j^a x_{ij} \leq (1 + \tau^a) \mu^a x_{ii} \qquad i \in V, \, a \in A$$

---

**Algorithm 3** Algorithm proposed in [19] based on IP formulation

---

**function** $SolveTDP_1$(type)
  **repeat**
    **if** $(type = IP)$ **then**
      Solve TDP$_{IPR_1}$
    **else**
      Solve TDP$_{QIPR}$
    **end if**
    Let $x^\star$ be the optimal solution of the relaxed formulation
    $S^\star := \emptyset$
    $nc := 0$
    **for all** $(i \in V : x_{ii}^\star = 1)$ **do**
      $nc := nc + 1$
      $m_{nc} := i$
      $C_{nc} := \emptyset$
      **for all** $(j \in V : x_{ij}^\star = 1)$ **do**
        $C_{nc} := C_{nc} \cup \{j\}$
      **end for**
      $S :=$FINDCONNECTEDCOMPONENTS$(G_{nc} = (C_{nc}, E(C_{nc}))$
      **for** $(u = 1, \ldots, |S|)$ **do**
        **if** $m_{nc} \notin S_u$ **then**
          $S^\star := S^\star \cup \{S_u\}$
        **end if**
      **end for**
    **end for**
    **if** $(S^\star \neq \emptyset)$ **then**
      **for all** $(S \in S^\star)$ **do**
        Add connectivity constraint associated to node subset $S$
        to the relaxed formulation
      **end for**
    **end if**
  **until** $S^\star = \emptyset$
  **return** $x^\star$
**end function**

---

$$x_{ij} \leq x_{ii} \qquad\qquad i, j \in V \qquad (5.5)$$

$$x_{ij} \in \{0, 1\} \qquad\qquad i, j \in V$$

Although constraints (5.5) are redundant for the integer programming formulation of the problem, they provide lower bounds of better quality for the linear programming relaxation of the problem. A procedure is used to find a feasible assignment of nodes to the $p$ medians obtained in the optimal solution of TP$_{IPR_2}$, that also satisfy connectivity constraints. Let $M$ be the set of medians in the optimal solution of TP$_{IPR_2}$. For this purpose the following assignment problem (AP) is considered,

$$(\text{AP}) \quad \min \quad \sum_{i \in M} \sum_{j \in V} d_{ij} x_{ij} \tag{5.6a}$$

$$\text{subject to} \quad \sum_{i \in M} x_{ij} = 1 \qquad\qquad\qquad j \in V \tag{5.6b}$$

$$\sum_{j \in V} w_j^a x_{ij} \geq (1 - \tau^a) \mu^a \qquad\qquad i \in M, \, a \in A$$
$$\tag{5.6c}$$

$$\sum_{j \in V} w_j^a x_{ij} \leq (1 + \tau^a) \mu^a \qquad\qquad i \in M, \, a \in A$$
$$\tag{5.6d}$$

$$\sum_{j \in \bigcup_{v \in S}(N^v \setminus S)} x_{ij} - \sum_{j \in S} x_{ij} \geq 1 - |S| \qquad i \in M,$$

$$S \subset V \setminus \left( N^i \cup \{i\} \right)$$
$$\tag{5.6e}$$

$$x_{ij} \in \{0, 1\} \qquad\qquad\qquad i \in M, \, j \in V$$
$$\tag{5.6f}$$

Since there is an exponential number of constraints (5.6e), we consider the following relaxation for the assignment subproblem:

$$(\text{Assignment}_{\text{Rel}}) \quad \min \quad \sum_{i \in M} \sum_{j \in V} d_{ij} x_{ij}$$

$$\text{subject to} \quad \sum_{i \in M} x_{ij} = 1 \qquad\qquad j \in V$$

$$\sum_{j \in V} w_j^a x_{ij} \geq (1 - \tau^a) \mu^a \qquad i \in V, \, a \in A$$

$$\sum_{j \in V} w_j^a x_{ij} \leq (1 + \tau^a) \mu^a \qquad i \in V, \, a \in A$$

$$x_{ij} \in \{0, 1\} \qquad\qquad\qquad i \in M, \, j \in V$$

where connectivity constraints (5.6e) are removed from the formulation. A similar iterative approach as the one used in Algorithm 3 is used to solve the assignment problem. In each iteration the relaxed assignment problem is solved, and all violated connectivity constraints are identified and added to the relaxed formulation until an

optimal assignment that satisfies connectivity constraints is found. Then, all violated
connectivity constraints identified to solve the assignment problem are also added
to TPD$_{IPR_2}$. As can be observed, the optimal solution of the assignment problem is
feasible for the TDP and can be used as an initial incumbent solution for TPD$_{IPR_2}$ in
the next iteration of the procedure. The use of an incumbent solution might help to
reduce the enumerative effort when solving the relaxed problem. Algorithms 4 and 5
detail the procedure used to find the optimal solution with the proposed procedure.

---

**Algorithm 4** Algorithm based on IP formulation TDP$_{IPR_2}$

---

    **function** $SolveTDP_2$
        Solve TDP$_{IPR_2}$
        **repeat**
            Let $x^\star$ be the optimal solution of the relaxed problem
            ASSIGNMENT($x^\star$, $S^\star$, $x_{best}$)
            **if** $(S^\star \neq \emptyset)$ **then**
                **for all** $(S \in S^\star)$ **do**
                    Add connectivity constraint associated to node subset $S$ to TDP$_{IPR_2}$
                    Solve TDP$_{IPR_2}$ using $x_{best}$ as an incumbent solution
                **end for**
            **end if**
        **until** $S^\star = \emptyset$
        **return** $x^\star$
    **end function**

---

Finally, we also propose a procedure to find feasible solutions to the TDP problem. To accomplish this, we incorporate a procedure to find feasible assignments
with respect to the connectivity constraints, to the Lagrangian relaxation approach
proposed in [5]. As mentioned in the previous section of this chapter, in this work,
the proposed Lagrangian relaxation is used to find lower bounds for the optimal
solution of the integer programming relaxation of TDP$_{IP}$ where the connectivity
constraints are dropped and the assignment constraints are relaxed in a Lagrangian
fashion. Also, in [5] a subgradient optimization algorithm is used to solve the
Lagrangian dual. In some iterations of this procedure (those in which the lower
bound improves) a primal heuristic is executed to obtain feasible solutions. Since in
[5], connectivity constraints are not considered in the problem formulation, we apply
the assignment procedure described in Algorithm 5 to the set of medians associated
with the best feasible solution found by the primal heuristic of the subgradient
optimization algorithm. This allows to find a feasible assignment to the problem
that also satisfies the connectivity constraints.

In the next section we will describe a series of computational experiments that
allow to evaluate and compare the different methods described in this section.

---

**Algorithm 5** Algorithm to find a feasible assignment of nodes to medians

---

**procedure** ASSIGNMENT($x^\star$, $S^\star$, $x_{best}$)
   $S^\star := \emptyset$
  **repeat**
     $T^\star := \emptyset$
     $nc := 0$
     **for all** $(i \in V : x_{ii}^\star = 1)$ **do**
        $nc := nc + 1$
        $m_{nc} := i$
        $C_{nc} := \emptyset$
        **for all** $(j \in V : x_{ij}^\star = 1)$ **do**
           $C_{nc} := C_{nc} \cup \{j\}$
        **end for**
        $S :=$ FINDCONNECTEDCOMPONENTS($G_{nc} = (C_{nc}, E(C_{nc}))$)
        **for** $(u = 1, \ldots, |S|)$ **do**
           **if** $m_{nc} \notin S_u$ **then**
              $T^\star := T^\star \cup \{S_u\}$
           **end if**
        **end for**
     **end for**
     **if** $T^\star \neq \emptyset$ **then**
        **for all** $T \in T^\star$ **do**
           Add connectivity constraint associated to node subset $T$ to
           $Assignment_{Rel}$
        **end for**
        $S^\star := S^\star \cup T^\star$
        Solve $Assignment_{Rel}$
        Let $x^\star$ be the optimal solution of $Assignment_{Rel}$
     **end if**
  **until** $T = \emptyset$
  $x_{best} := x^\star$
**end procedure**

---

## 5.4 Computational Results

The purpose of this section is to present the results obtained from the computational tests carried out with the methods described in the previous section. These results will allow to evaluate and compare the behavior of each one of them. In particular, we provide results concerning the quality of the lower bounds obtained with different relaxations of the problem, the quality of the upper bounds obtained with methods that do not guarantee solution optimality, and the computational effort required by the exact methods described, for the territory design problem studied. The considered methods are the following:

- The exact procedure using the $\text{TDP}_{\text{IP}_1}$ formulation of the problem proposed in [19].
- The exact procedure using the $\text{TDP}_{\text{IP}_2}$ formulation of the problem proposed in this work.
- The procedure to find feasible solutions of the problem that uses the $\text{TDP}_{\text{QIP}}$ formulation of the problem proposed in [19].
- The procedure to find feasible solutions proposed in this chapter that uses the Lagrangian relaxation proposed in [5] and finds a feasible assignment of BUs to medians by solving an assignment problem.

With the exception of the method based on the quadratic programming formulation proposed in [19], all methodologies were coded in FICO XPRESS optimization suite 8.2 and executed in a computer with an Intel Xeon processor with an E3-1220 @3.1 GHz CPU and with 8 GB in RAM. The results of the method that uses the quadratic formulation were taken from [19] and, therefore, the execution times of this method cannot be compared with the execution times of the remaining methods since they correspond to a different computer. These formulations were solved by DICOPT, which is a non-linear mixed integer program solver.

We use the same set of test instances of the territory design problem used in [19] and in [5]. These test instances were generated by Salazar-Aguilar et al. [19] using the instance generator developed by Ríos-Mercado and Fernández in [17] and are publicly available at http://yalma.fime.uanl.mx/~roger/ftp/tdp. This instance generator is based on real-world data provided by industry and contains 80 test instances divided into five subsets of instances according to their size ($|V|$, $p$):

- (60,4): contains 20 instances with 60 BUs and $p = 4$.
- (80,5): contains 20 instances with 80 BUs and $p = 5$.
- (100,6): contains 20 instances with 100 BUs and $p = 6$.
- (150,8): contains 10 instances with 150 BUs and $p = 8$.
- (200,11): contains 10 instances with 200 BUs and $p = 11$.

We first comment on the quality of the lower bounds obtained with the different methods described in Sect. 5.3. Tables 5.1, 5.2, 5.3, 5.4, and 5.5 show the lower bounds obtained with the linear programming relaxations of $\text{TDP}_{\text{IPR}_1}$ and $\text{TDP}_{\text{IPR}_2}$, and the lower bounds provided by the Lagrangian relaxation approach proposed by Díaz and Luna [5] for groups (60, 4) to (200, 11), respectively. The data in these tables are:

- Column 1: Data instance name (Instance name).
- Column 2: Optimal solution of the instance (Optimal Value).
- Column 3: Lower bound provided by the linear programming relaxation of $\text{TDP}_{\text{IP}_1}$ (LB LP $\text{TDP}_{\text{IPR}_1}$).
- Column 4: Lower bound provided by the linear programming relaxation of $\text{TDP}_{\text{IP}_2}$ (LB LP $\text{TDP}_{\text{IPR}_2}$).
- Column 5: Lower bound provided by the Lagrangian relaxation approach (LB LR).

- Column 6: percentage deviation of the lower bound obtained with the linear programming relaxation of $TDP_{IPR_1}$ with respect to the optimal solution (gap $TDP_{IPR_1}$).
- Column 7: percentage deviation of the lower bound obtained with the linear programming relaxation of $TDP_{IPR_2}$ with respect to the optimal solution (gap $TDP_{IPR_2}$).
- Column 8: percentage deviation of the lower bound obtained with the Lagrangian relaxation with respect to the optimal solution (gap LR).

As can be observed in Tables 5.1, 5.2, 5.3, 5.4, and 5.5, the method that provides the best lower bounds is the Lagrangian relaxation proposed in [5], followed by the lower bounds obtained with the linear programming relaxation of $TDP_{IPR_2}$. However, lower bounds provided by the linear relaxation of $TDP_{IPR_1}$ are weak, since on average, the percentage gap of the linear programming relaxation of $TDP_{IPR_1}$ with respect to the optimal solutions are 185.85%, 196.51%, 206.56%, 212.06%, and 204.00% for instance groups (60, 4), (80, 5), (100, 6), (150, 8), and (200, 11), respectively. On the contrary, the average percentage gap of the linear programming relaxation of $TDP_{IPR_2}$ with respect to the optimal solution are, respectively, 1.07%, 1.64%, 0.93%, 1.02%, and 1.15% for instance groups (60, 4), (80, 5), (100, 6), (150, 8), and (200, 11), and the average percentage gap of the Lagrangian relaxation with

**Table 5.1** Lower bounds for (60, 4) instances

| Instance name | Optimal value | LB LP $TDP_{IPR_1}$ | LB LP $TDP_{IPR_2}$ | LB LR | Gap $TDP_{IPR_1}$ | Gap $TDP_{IPR_2}$ | Gap $LR$ |
|---|---|---|---|---|---|---|---|
| 2DU60-05-1 | 5305.6 | 2030.6 | 5272.0 | 5297.4 | 161.28% | 0.64% | 0.15% |
| 2DU60-05-2 | 5451.7 | 1898.5 | 5370.1 | 5439.5 | 187.16% | 1.52% | 0.22% |
| 2DU60-05-3 | 5507.9 | 1806.9 | 5437.0 | 5497.7 | 204.82% | 1.30% | 0.19% |
| 2DU60-05-4 | 5935.7 | 1952.3 | 5857.7 | 5935.3 | 204.03% | 1.33% | 0.01% |
| 2DU60-05-5 | 5303.2 | 1610.5 | 5255.3 | 5303.0 | 229.30% | 0.91% | 0.00% |
| 2DU60-05-6 | 5253.9 | 1757.9 | 5163.9 | 5228.6 | 198.88% | 1.74% | 0.48% |
| 2DU60-05-7 | 5460.2 | 1997.1 | 5414.8 | 5459.5 | 173.41% | 0.84% | 0.01% |
| 2DU60-05-8 | 5310.0 | 2110.9 | 5260.9 | 5309.7 | 151.55% | 0.93% | 0.01% |
| 2DU60-05-9 | 5224.5 | 2066.1 | 5178.1 | 5222.4 | 152.87% | 0.90% | 0.04% |
| 2DU60-05-10 | 5350.2 | 2087.3 | 5313.8 | 5348.1 | 156.32% | 0.68% | 0.04% |
| 2DU60-05-11 | 5150.9 | 1901.4 | 5114.3 | 5142.1 | 170.90% | 0.72% | 0.17% |
| 2DU60-05-12 | 5597.5 | 2026.4 | 5527.2 | 5587.0 | 176.23% | 1.27% | 0.19% |
| 2DU60-05-13 | 5732.0 | 2097.1 | 5662.6 | 5731.2 | 173.33% | 1.23% | 0.01% |
| 2DU60-05-14 | 5463.0 | 2294.9 | 5403.1 | 5461.9 | 138.05% | 1.11% | 0.02% |
| 2DU60-05-15 | 5332.8 | 1932.4 | 5266.8 | 5331.8 | 175.97% | 1.25% | 0.02% |
| 2DU60-05-16 | 5399.5 | 1866.7 | 5330.5 | 5368.7 | 189.25% | 1.29% | 0.57% |
| 2DU60-05-17 | 5602.9 | 2194.2 | 5546.2 | 5602.0 | 155.35% | 1.02% | 0.02% |
| 2DU60-05-18 | 5774.0 | 1786.3 | 5743.5 | 5773.8 | 223.24% | 0.53% | 0.00% |
| 2DU60-05-19 | 5543.5 | 1512.7 | 5464.7 | 5526.4 | 266.47% | 1.44% | 0.31% |
| 2DU60-05-20 | 5767.5 | 1754.9 | 5727.2 | 5767.2 | 228.65% | 0.70% | 0.01% |

**Table 5.2** Lower bounds for (80, 5) instances

| Instance name | Optimal value | LB LP $TDP_{IPR_1}$ | LB LP $TDP_{IPR_2}$ | LB LR | Gap $TDP_{IPR_1}$ | Gap $TDP_{IPR_2}$ | Gap $LR$ |
|---|---|---|---|---|---|---|---|
| 2DU80-05-1 | 6600.6 | 2268.1 | 6366.1 | 6405.8 | 191.02% | 3.68% | 3.04% |
| 2DU80-05-2 | 6408.8 | 2223.8 | 6347.4 | 6408.4 | 188.19% | 0.97% | 0.01% |
| 2DU80-05-3 | 6958.1 | 2402.6 | 6903.3 | 6957.4 | 189.61% | 0.79% | 0.01% |
| 2DU80-05-4 | 6900.2 | 2249.2 | 6698.8 | 6743.4 | 206.78% | 3.01% | 2.32% |
| 2DU80-05-5 | 6280.6 | 2087.1 | 6155.6 | 6234.4 | 200.93% | 2.03% | 0.74% |
| 2DU80-05-6 | 6521.1 | 2112.5 | 6385.7 | 6493.3 | 208.69% | 2.12% | 0.43% |
| 2DU80-05-7 | 6456.0 | 2125.7 | 6356.2 | 6415.9 | 203.71% | 1.57% | 0.62% |
| 2DU80-05-8 | 6680.3 | 2102.0 | 6545.7 | 6639.5 | 217.81% | 2.06% | 0.61% |
| 2DU80-05-9 | 6650.2 | 2314.9 | 6497.6 | 6513.3 | 187.28% | 2.35% | 2.10% |
| 2DU80-05-10 | 6534.8 | 2483.5 | 6366.5 | 6401.4 | 163.13% | 2.64% | 2.08% |
| 2DU80-05-11 | 6539.6 | 2281.5 | 6451.9 | 6522.5 | 186.63% | 1.36% | 0.26% |
| 2DU80-05-12 | 6704.0 | 2174.6 | 6640.1 | 6694.3 | 208.29% | 0.96% | 0.14% |
| 2DU80-05-13 | 6285.7 | 2136.4 | 6232.3 | 6262.2 | 194.22% | 0.86% | 0.38% |
| 2DU80-05-14 | 6615.8 | 2263.2 | 6501.4 | 6559.2 | 192.32% | 1.76% | 0.86% |
| 2DU80-05-15 | 6990.4 | 2364.5 | 6930.9 | 6981.9 | 195.64% | 0.86% | 0.12% |
| 2DU80-05-16 | 6391.7 | 2120.4 | 6319.5 | 6372.7 | 201.44% | 1.14% | 0.30% |
| 2DU80-05-17 | 6766.0 | 2490.2 | 6740.9 | 6754.7 | 171.70% | 0.37% | 0.17% |
| 2DU80-05-18 | 6808.5 | 2248.6 | 6721.0 | 6772.1 | 202.79% | 1.30% | 0.54% |
| 2DU80-05-19 | 6643.2 | 2044.9 | 6528.8 | 6598.0 | 224.87% | 1.75% | 0.68% |
| 2DU80-05-20 | 6873.6 | 2328.9 | 6784.7 | 6831.8 | 195.15% | 1.31% | 0.61% |

respect to the optimal solution are, respectively, 0.12%, 0.8%, 0.53%, 0.58%, and 0.68% for instance groups (60, 4), (80, 5), (100, 6), (150, 8), and (200, 11).

To evaluate the exact methods, in Tables 5.6, 5.7, 5.8, 5.9, and 5.10 we show the results obtained with the methods described in Sect. 5.3: the exact method which uses the linear integer programming relaxation $TDP_{IPR_1}$, and the exact method that uses the linear integer programming relaxation $TDP_{IPR_2}$. In these Tables, the first two columns show, respectively, the instance name, and the value of the optimal solution. Then, the next two columns show, respectively, the CPU times (in seconds) of the method that uses $TDP_{IPR_1}$, and the method that uses $TDP_{IPR_2}$. As can be observed, the method that uses the integer programming relaxation $TDP_{IPR_2}$, in general, obtains the optimal solution of the instances in less time than that required by the method that uses the integer linear programming relaxation $TDP_{IPR_1}$. The average times needed to solve, respectively, the (60, 4), (80, 5), (100, 6), (150, 8), and (200, 11) instances are 4.32, 32.25, 51.26, 351.94, and 1548.59 for the method that uses $TDP_{IPR_1}$, and 0.90, 8.34, 12.61, 34.31, and 170.72 for the method that uses $TDP_{IPR_2}$. These times can be clearly appreciated in Fig. 5.1. The observed difference in the CPU times required by the method based on relaxation $TDP_{IPR_2}$ may be due to the combination of two factors. First, the lower bound of linear programming relaxation of $TDP_{IPR_2}$ are much better than those obtained with the linear programming relaxation of $TDP_{IPR_1}$. Second, in the instances in which it is

**Table 5.3** Lower bounds for (100, 6) instances

| Instance name | Optimal value | LB LP $TDP_{IPR_1}$ | LB LP $TDP_{IPR_2}$ | LB LR | Gap $TDP_{IPR_1}$ | Gap $TDP_{IPR_2}$ | Gap $LR$ |
|---|---|---|---|---|---|---|---|
| 2DU100-05-1 | 7370.1 | 2341.7 | 7228.2 | 7291.2 | 214.74% | 1.96% | 1.08% |
| 2DU100-05-2 | 7278.5 | 2355.6 | 7198.0 | 7216.2 | 208.98% | 1.12% | 0.86% |
| 2DU100-05-3 | 7512.3 | 2354.6 | 7448.9 | 7465.1 | 219.05% | 0.85% | 0.63% |
| 2DU100-05-4 | 7581.6 | 2510.0 | 7520.8 | 7571.3 | 202.05% | 0.81% | 0.14% |
| 2DU100-05-5 | 7609.5 | 2481.8 | 7523.2 | 7549.8 | 206.61% | 1.15% | 0.79% |
| 2DU100-05-6 | 7243.0 | 2429.0 | 7182.5 | 7205.5 | 198.18% | 0.84% | 0.52% |
| 2DU100-05-7 | 7432.7 | 2456.4 | 7415.5 | 7432.2 | 202.59% | 0.23% | 0.01% |
| 2DU100-05-8 | 7052.9 | 2123.6 | 7029.8 | 7051.5 | 232.12% | 0.33% | 0.02% |
| 2DU100-05-9 | 7181.5 | 2471.1 | 7111.2 | 7151.6 | 190.62% | 0.99% | 0.42% |
| 2DU100-05-10 | 7432.9 | 2428.6 | 7388.2 | 7400.3 | 206.05% | 0.61% | 0.44% |
| 2DU100-05-11 | 6829.5 | 2552.4 | 6783.2 | 6827.8 | 167.57% | 0.68% | 0.02% |
| 2DU100-05-12 | 7461.2 | 2211.0 | 7444.3 | 7449.4 | 237.46% | 0.23% | 0.16% |
| 2DU100-05-13 | 7061.6 | 2601.4 | 7045.9 | 7058.8 | 171.45% | 0.22% | 0.04% |
| 2DU100-05-14 | 7825.6 | 2373.6 | 7818.5 | 7824.3 | 229.69% | 0.09% | 0.02% |
| 2DU100-05-15 | 7158.7 | 2353.2 | 7101.1 | 7124.2 | 204.21% | 0.81% | 0.49% |
| 2DU100-05-16 | 7653.2 | 2462.0 | 7529.9 | 7550.8 | 210.85% | 1.64% | 1.36% |
| 2DU100-05-17 | 6880.5 | 2045.4 | 6748.6 | 6765.4 | 236.40% | 1.95% | 1.70% |
| 2DU100-05-18 | 7438.5 | 2546.7 | 7336.2 | 7388.0 | 192.09% | 1.39% | 0.68% |
| 2DU100-05-19 | 7238.1 | 2369.9 | 7135.3 | 7210.0 | 205.42% | 1.44% | 0.39% |
| 2DU100-05-20 | 7590.1 | 2573.0 | 7503.4 | 7534.1 | 194.99% | 1.16% | 0.74% |

**Table 5.4** Lower bounds for (150, 8) instances

| Instance name | Optimal value | LB LP $TDP_{IPR_1}$ | LB LP $TDP_{IPR_2}$ | LB LR | Gap $TDP_{IPR_1}$ | Gap $TDP_{IPR_2}$ | Gap $LR$ |
|---|---|---|---|---|---|---|---|
| DU150-05-1 | 9511.8 | 3076.7 | 9427.0 | 9460.8 | 209.15% | 0.90% | 0.54% |
| DU150-05-2 | 9400.5 | 2939.1 | 9265.0 | 9335.8 | 219.85% | 1.46% | 0.69% |
| DU150-05-3 | 9134.6 | 2888.6 | 9089.7 | 9113.6 | 216.23% | 0.49% | 0.23% |
| DU150-05-4 | 9359.0 | 3114.6 | 9319.9 | 9342.2 | 200.49% | 0.42% | 0.18% |
| DU150-05-5 | 9506.6 | 2959.3 | 9456.0 | 9484.4 | 221.25% | 0.53% | 0.23% |
| DU150-05-6 | 9039.1 | 2960.2 | 8972.3 | 9038.7 | 205.35% | 0.74% | 0.00% |
| DU150-05-7 | 9854.7 | 2997.4 | 9702.1 | 9745.0 | 228.77% | 1.57% | 1.13% |
| DU150-05-8 | 9199.3 | 3173.7 | 8960.9 | 9014.0 | 189.86% | 2.66% | 2.06% |
| DU150-05-9 | 9670.9 | 2862.9 | 9561.8 | 9617.4 | 237.80% | 1.14% | 0.56% |
| DU150-05-10 | 9570.6 | 3279.7 | 9540.8 | 9555.8 | 191.81% | 0.31% | 0.15% |

**Table 5.5** Lower bounds for (200, 11) instances

| Instance name | Optimal value | LB LP TDP$_{IPR_1}$ | LB LP TDP$_{IPR_2}$ | LB LR | Gap TDP$_{IPR_1}$ | Gap TDP$_{IPR_2}$ | Gap $LR$ |
|---|---|---|---|---|---|---|---|
| DU200-05-1 | 10,422.0 | 3571.9 | 10,331.5 | 10,391.1 | 191.77% | 0.88% | 0.30% |
| DU200-05-2 | 10,639.8 | 3516.8 | 10,528.8 | 10,576.0 | 202.55% | 1.05% | 0.60% |
| DU200-05-3 | 10,837.9 | 3633.6 | 10,759.5 | 10,807.1 | 198.27% | 0.73% | 0.28% |
| DU200-05-4 | 11,124.9 | 3546.5 | 10,970.0 | 11,047.0 | 213.69% | 1.41% | 0.71% |
| DU200-05-5 | 10,874.5 | 3478.9 | 10,727.5 | 10,766.5 | 212.59% | 1.37% | 1.00% |
| DU200-05-6 | 10,492.2 | 3731.0 | 10,343.1 | 10,390.5 | 181.22% | 1.44% | 0.98% |
| DU200-05-7 | 11,020.9 | 3525.6 | 10,848.7 | 10,889.2 | 212.60% | 1.59% | 1.21% |
| DU200-05-8 | 10,650.7 | 3484.5 | 10,531.9 | 10,580.5 | 205.66% | 1.13% | 0.66% |
| DU200-05-9 | 11,431.3 | 3583.7 | 11,281.5 | 11,342.7 | 218.98% | 1.33% | 0.78% |
| DU200-05-10 | 11,039.5 | 3647.2 | 10,973.0 | 11,006.6 | 202.68% | 0.61% | 0.30% |

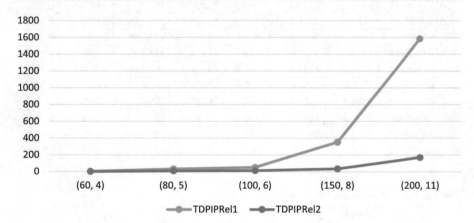

**Fig. 5.1** CPU time (in seconds) exact methods

required to add cuts associated with violated connectivity constraints, an incumbent solution is at hand when re-optimizing TDP$_{IPR_2}$, that might help to reduce the enumerative effort.

Finally, the results obtained with the methods that provide feasible solutions (not necessarily optimal), are shown in Tables 5.11, 5.12, 5.13, 5.14, and 5.15. Since in [19] only the results associated with the relaxation of the integer quadratic formulation of the problem TPD$_{QIPR}$ are reported for the sets of instances (60, 4), (150, 8), and (200, 11), Tables 5.12 and 5.13 show only the results of the Lagrangian relaxation-based approach for the instance sets (80, 5) and (100, 6). The values reported in these Tables are: Column (1) shows the instance name, Column(2) depicts the optimal solution value, Columns (3), (4), and (5) display, respectively, the value of the best upper bound, the CPU time in seconds, and the gap of the upper bound with respect to the optimal solution, for the Lagrangian relaxation approach. Finally, Columns (6), (7), and (8) show, respectively, the value of the best upper bound, the CPU time in seconds, and the gap of the upper bound with respect

**Table 5.6** CPU time (in seconds) of exact methods for (60, 4) instances

| Instance name | Optimal value | CPU time $TDP_{IPR_1}$ | CPU time $TDP_{IPR_2}$ |
|---|---|---|---|
| 2DU60-05-1 | 5305.6 | 4.03 | 0.55 |
| 2DU60-05-2 | 5451.7 | 4.16 | 0.92 |
| 2DU60-05-3 | 5507.9 | 5.76 | 1.65 |
| 2DU60-05-4 | 5935.7 | 3.41 | 0.35 |
| 2DU60-05-5 | 5303.2 | 2.20 | 0.38 |
| 2DU60-05-6 | 5253.9 | 5.69 | 3.18 |
| 2DU60-05-7 | 5460.2 | 2.84 | 0.34 |
| 2DU60-05-8 | 5310.0 | 2.19 | 0.42 |
| 2DU60-05-9 | 5224.5 | 1.83 | 0.31 |
| 2DU60-05-10 | 5350.2 | 2.58 | 0.35 |
| 2DU60-05-11 | 5150.9 | 2.80 | 0.60 |
| 2DU60-05-12 | 5597.5 | 7.60 | 1.13 |
| 2DU60-05-13 | 5732.0 | 3.32 | 0.52 |
| 2DU60-05-14 | 5463.0 | 3.51 | 0.67 |
| 2DU60-05-15 | 5332.8 | 4.01 | 0.69 |
| 2DU60-05-16 | 5399.5 | 13.91 | 2.43 |
| 2DU60-05-17 | 5602.9 | 1.88 | 0.39 |
| 2DU60-05-18 | 5774.0 | 3.76 | 0.37 |
| 2DU60-05-19 | 5543.5 | 6.66 | 2.28 |
| 2DU60-05-20 | 5767.5 | 3.32 | 0.41 |

**Table 5.7** CPU time (in seconds) of exact methods for (80, 5) instances

| Instance name | Optimal value | CPU time $TDP_{IPR_1}$ | CPU time $TDP_{IPR_2}$ |
|---|---|---|---|
| 2DU80-05-1 | 6600.6 | 87.13 | 20.50 |
| 2DU80-05-2 | 6408.8 | 6.87 | 0.59 |
| 2DU80-05-3 | 6958.1 | 12.76 | 1.98 |
| 2DU80-05-4 | 6900.2 | 108.77 | 22.98 |
| 2DU80-05-5 | 6280.6 | 25.35 | 7.32 |
| 2DU80-05-6 | 6521.1 | 23.95 | 6.31 |
| 2DU80-05-7 | 6456.0 | 33.17 | 17.27 |
| 2DU80-05-8 | 6680.3 | 28.82 | 6.30 |
| 2DU80-05-9 | 6650.2 | 53.94 | 13.78 |
| 2DU80-05-10 | 6534.8 | 68.00 | 14.67 |
| 2DU80-05-11 | 6539.6 | 8.60 | 2.75 |
| 2DU80-05-12 | 6704.0 | 9.72 | 1.40 |
| 2DU80-05-13 | 6285.7 | 16.79 | 1.78 |
| 2DU80-05-14 | 6615.8 | 39.62 | 8.80 |
| 2DU80-05-15 | 6990.4 | 15.19 | 2.08 |
| 2DU80-05-16 | 6391.7 | 14.97 | 4.07 |
| 2DU80-05-17 | 6766.0 | 19.10 | 2.94 |
| 2DU80-05-18 | 6808.5 | 17.50 | 13.85 |
| 2DU80-05-19 | 6643.2 | 22.57 | 6.43 |
| 2DU80-05-20 | 6873.6 | 36.04 | 11.00 |

**Table 5.8** CPU time (in seconds) of exact methods for (100, 6) instances

| Instance name | Optimal value | CPU time $TDP_{IPR_1}$ | CPU time $TDP_{IPR_2}$ |
|---|---|---|---|
| 2DU100-05-1 | 7370.1 | 118.51 | 24.70 |
| 2DU100-05-2 | 7278.5 | 57.98 | 21.85 |
| 2DU100-05-3 | 7512.3 | 41.21 | 9.58 |
| 2DU100-05-4 | 7581.6 | 13.48 | 2.02 |
| 2DU100-05-5 | 7609.5 | 146.36 | 14.10 |
| 2DU100-05-6 | 7243.0 | 65.58 | 7.79 |
| 2DU100-05-7 | 7432.7 | 12.92 | 0.78 |
| 2DU100-05-8 | 7052.9 | 11.49 | 1.72 |
| 2DU100-05-9 | 7181.5 | 44.89 | 9.37 |
| 2DU100-05-10 | 7432.9 | 34.29 | 8.58 |
| 2DU100-05-11 | 6829.5 | 18.22 | 1.10 |
| 2DU100-05-12 | 7461.2 | 16.87 | 0.97 |
| 2DU100-05-13 | 7061.6 | 11.04 | 0.77 |
| 2DU100-05-14 | 7825.6 | 14.44 | 0.74 |
| 2DU100-05-15 | 7158.7 | 61.98 | 17.75 |
| 2DU100-05-16 | 7653.2 | 94.82 | 59.07 |
| 2DU100-05-17 | 6880.5 | 124.40 | 30.70 |
| 2DU100-05-18 | 7438.5 | 62.72 | 9.34 |
| 2DU100-05-19 | 7238.1 | 22.84 | 13.72 |
| 2DU100-05-20 | 7590.1 | 79.02 | 17.56 |

**Table 5.9** CPU time (in seconds) of exact methods for (150, 8) instances

| Instance name | Optimal value | CPU time $TDP_{IPR_1}$ | CPU time $TDP_{IPR_2}$ |
|---|---|---|---|
| DU150-05-1 | 9511.8 | 170.60 | 43.84 |
| DU150-05-2 | 9400.5 | 373.44 | 133.94 |
| DU150-05-3 | 9134.6 | 93.58 | 12.99 |
| DU150-05-4 | 9359.0 | 116.21 | 19.89 |
| DU150-05-5 | 9506.6 | 120.94 | 7.74 |
| DU150-05-6 | 9039.1 | 55.03 | 6.57 |
| DU150-05-7 | 9854.7 | 1295.88 | 112.94 |
| DU150-05-8 | 9199.3 | 3666.36 | 158.65 |
| DU150-05-9 | 9670.9 | 288.51 | 33.41 |
| DU150-05-10 | 9570.6 | 102.87 | 4.52 |

to the optimal solution, to the method that uses the relaxation for the quadratic integer programming formulation. We observed some inconsistencies in the values reported in [19] for some instances of size (150, 8) and (200, 11). For example, DU150-05-3, the optimal solution of $TDP_{IPR_1}$ is 9134.6 and the optimal solution reported in Salazar is 9125.6 for $TDP_{IP}$ and 9130 for $TDP_{QIP}$. The values reported in the table with n.a. correspond to test instance DU150-05-3 where the optimal solution reported is smaller than the optimum of the instance. As can be observed, the Lagrangian relaxation provides good upper bounds for the optimal solution of

**Table 5.10** CPU time (in seconds) of exact methods for (200, 11) instances

| Instance name | Optimal value | CPU time $TDP_{IPR_1}$ | CPU time $TDP_{IPR_2}$ |
|---|---|---|---|
| DU200-05-1 | 10,422.0 | 534.02 | 41.35 |
| DU200-05-2 | 10,639.8 | 2075.07 | 313.51 |
| DU200-05-3 | 10,837.9 | 415.29 | 43.17 |
| DU200-05-4 | 11,124.9 | 4075.79 | 383.36 |
| DU200-05-5 | 10,874.5 | 1588.46 | 203.92 |
| DU200-05-6 | 10,492.2 | 5632.98 | 194.01 |
| DU200-05-7 | 11,020.9 | 4456.03 | 964.35 |
| DU200-05-8 | 10,650.7 | 532.78 | 107.08 |
| DU200-05-9 | 11,431.3 | 4513.34 | 563.21 |
| DU200-05-10 | 11,039.5 | 522.67 | 65.98 |

**Table 5.11** Methods to find feasible solutions for (60, 4) instances

| Instance name | Optimal value | Lagrangian relaxation | | | $TDP_{QIPR}$ | | |
|---|---|---|---|---|---|---|---|
| | | UB | CPU time | gap | UB | CPU time | gap |
| 2DU60-05-1 | 5305.6 | 5305.6 | 41.80 | 0.00% | 5305.6 | 2.00 | 0.00% |
| 2DU60-05-2 | 5451.7 | 5463.4 | 61.54 | 0.22% | 5463.0 | 2.00 | 0.21% |
| 2DU60-05-3 | 5507.9 | 5507.9 | 106.75 | 0.00% | 5553.0 | 2.00 | 0.82% |
| 2DU60-05-4 | 5935.7 | 5935.7 | 5.54 | 0.00% | 6114.0 | 6.00 | 3.00% |
| 2DU60-05-5 | 5303.2 | 5303.2 | 3.54 | 0.00% | 5303.2 | 2.00 | 0.00% |
| 2DU60-05-6 | 5253.9 | 5257.9 | 48.33 | 0.08% | 5280.0 | 3.00 | 0.50% |
| 2DU60-05-7 | 5460.2 | 5460.2 | 10.99 | 0.00% | 5855.0 | 3.00 | 7.23% |
| 2DU60-05-8 | 5310.0 | 5310.0 | 33.59 | 0.00% | 5314.0 | 2.00 | 0.08% |
| 2DU60-05-9 | 5224.5 | 5224.5 | 13.52 | 0.00% | 5224.5 | 3.00 | 0.00% |
| 2DU60-05-10 | 5350.2 | 5350.2 | 16.70 | 0.00% | 6140.0 | 2.00 | 14.76% |
| 2DU60-05-11 | 5150.9 | 5150.9 | 54.84 | 0.00% | 5152.0 | 2.00 | 0.02% |
| 2DU60-05-12 | 5597.5 | 5597.5 | 89.30 | 0.00% | 5705.0 | 2.00 | 1.92% |
| 2DU60-05-13 | 5732.0 | 5732.0 | 24.38 | 0.00% | 5732.0 | 3.00 | 0.00% |
| 2DU60-05-14 | 5463.0 | 5463.0 | 41.71 | 0.00% | 5869.0 | 2.00 | 7.43% |
| 2DU60-05-15 | 5332.8 | 5332.8 | 78.26 | 0.00% | 5759.0 | 2.00 | 7.99% |
| 2DU60-05-16 | 5399.5 | 5399.5 | 75.94 | 0.00% | 5499.0 | 2.00 | 1.84% |
| 2DU60-05-17 | 5602.9 | 5602.9 | 8.90 | 0.00% | 5602.9 | 2.00 | 0.00% |
| 2DU60-05-18 | 5774.0 | 5774.0 | 11.83 | 0.00% | 6299.0 | 4.00 | 9.09% |
| 2DU60-05-19 | 5543.5 | 5564.0 | 42.52 | 0.37% | 5543.5 | 2.00 | 0.00% |
| 2DU60-05-20 | 5767.5 | 5767.5 | 6.07 | 0.00% | 5767.5 | 2.00 | 0.00% |

the problem. The average relative gaps for this method are 0.03%, 0.05%, 0.04%, 0.05%, and 0.09% for the (60, 4), (80, 5), (100, 6), (150, 8), and (200, 11) instances, respectively. Also, the optimal solution is obtained in 53 out of the 80 test instances. According to the result shown in Tables 5.11, 5.14, and 5.15, the procedure that uses $TDP_{QIPR}$ relaxation obtains the optimal solution 8 out of the 40 test instances reported in [19]. Average relative gaps are 2.74%, 3.10%, and 5.07% for the (60, 4), (150, 8), and (200, 11) instances, respectively. In Fig. 5.2, it is observed that CPU

**Table 5.12** Methods to find feasible solutions for (80, 5) instances

| Instance name | Optimal value | Lagrangian relaxation | | |
|---|---|---|---|---|
| | | UB | CPU time | gap |
| 2DU80-05-1 | 6600.6 | 6601.1 | 91.60 | 0.01% |
| 2DU80-05-2 | 6408.8 | 6408.8 | 10.51 | 0.00% |
| 2DU80-05-3 | 6958.1 | 6958.1 | 68.89 | 0.00% |
| 2DU80-05-4 | 6900.2 | 6900.2 | 114.39 | 0.00% |
| 2DU80-05-5 | 6280.6 | 6281.6 | 110.03 | 0.02% |
| 2DU80-05-6 | 6521.1 | 6521.1 | 136.36 | 0.00% |
| 2DU80-05-7 | 6456.0 | 6456.0 | 81.26 | 0.00% |
| 2DU80-05-8 | 6680.3 | 6680.3 | 118.74 | 0.00% |
| 2DU80-05-9 | 6650.2 | 6650.2 | 103.26 | 0.00% |
| 2DU80-05-10 | 6534.8 | 6534.8 | 125.32 | 0.00% |
| 2DU80-05-11 | 6539.6 | 6559.9 | 50.47 | 0.31% |
| 2DU80-05-12 | 6704.0 | 6704.0 | 81.69 | 0.00% |
| 2DU80-05-13 | 6285.7 | 6287.2 | 57.08 | 0.02% |
| 2DU80-05-14 | 6615.8 | 6615.8 | 102.43 | 0.00% |
| 2DU80-05-15 | 6990.4 | 6990.4 | 61.65 | 0.00% |
| 2DU80-05-16 | 6391.7 | 6391.7 | 110.29 | 0.00% |
| 2DU80-05-17 | 6766.0 | 6771.0 | 60.47 | 0.07% |
| 2DU80-05-18 | 6808.5 | 6808.5 | 105.66 | 0.00% |
| 2DU80-05-19 | 6643.2 | 6643.2 | 103.31 | 0.00% |
| 2DU80-05-20 | 6873.6 | 6915.6 | 116.03 | 0.61% |

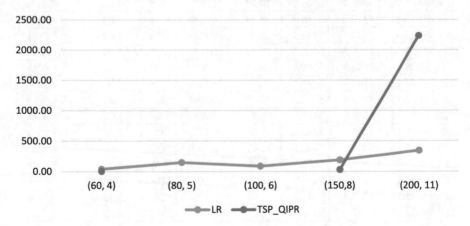

**Fig. 5.2** CPU time (in seconds) for Lagrangian relaxation and TDP$_{QIPR}$

times are very small for TDP$_{QIPR}$ relaxation for instances up to 150 BUs an $p = 8$, but they have a significant increase for test instances with 200 BUs and $p = 11$. On the contrary, the growth of CPU time as the size of the instances increases is much more moderate for Lagrangian relaxation.

**Table 5.13** Methods to find feasible solutions for (100, 5) instances

| Instance name | Optimal value | Lagrangian relaxation | | |
|---|---|---|---|---|
| | | UB | CPU time | gap |
| 2DU100-05-1 | 7370.1 | 7370.1 | 114.06 | 0.00% |
| 2DU100-05-2 | 7278.5 | 7278.5 | 101.51 | 0.00% |
| 2DU100-05-3 | 7512.3 | 7512.3 | 83.41 | 0.00% |
| 2DU100-05-4 | 7581.6 | 7581.6 | 137.69 | 0.00% |
| 2DU100-05-5 | 7609.5 | 7609.5 | 134.33 | 0.00% |
| 2DU100-05-6 | 7243.0 | 7243.0 | 97.69 | 0.00% |
| 2DU100-05-7 | 7432.7 | 7432.7 | 23.32 | 0.00% |
| 2DU100-05-8 | 7052.9 | 7052.9 | 71.33 | 0.00% |
| 2DU100-05-9 | 7181.5 | 7181.5 | 107.97 | 0.00% |
| 2DU100-05-10 | 7432.9 | 7442.0 | 72.54 | 0.12% |
| 2DU100-05-11 | 6829.5 | 6829.5 | 59.90 | 0.00% |
| 2DU100-05-12 | 7461.2 | 7461.2 | 18.32 | 0.00% |
| 2DU100-05-13 | 7061.6 | 7061.6 | 23.64 | 0.00% |
| 2DU100-05-14 | 7825.6 | 7825.6 | 16.76 | 0.00% |
| 2DU100-05-15 | 7158.7 | 7170.4 | 99.78 | 0.16% |
| 2DU100-05-16 | 7653.2 | 7657.8 | 121.97 | 0.06% |
| 2DU100-05-17 | 6880.5 | 6890.6 | 107.37 | 0.15% |
| 2DU100-05-18 | 7438.5 | 7438.5 | 90.23 | 0.00% |
| 2DU100-05-19 | 7238.1 | 7238.1 | 165.35 | 0.00% |
| 2DU100-05-20 | 7590.1 | 7612.8 | 84.00 | 0.30% |

**Table 5.14** Methods to find feasible solutions for (150, 8) instances

| Instance name | Optimal value | Lagrangian relaxation | | | $TDP_{QIPR}$ | | |
|---|---|---|---|---|---|---|---|
| | | UB | CPU time | gap | UB | CPU time | gap |
| DU150-05-1 | 9511.8 | 9511.8 | 177.35 | 0.00% | 9979.0 | 9.00 | 4.91% |
| DU150-05-2 | 9400.5 | 9400.5 | 287.86 | 0.00% | 9509.0 | 29.00 | 1.15% |
| DU150-05-3 | 9134.6 | 9135.6 | 182.41 | 0.01% | n.a. | n.a. | n.a.% |
| DU150-05-4 | 9359.0 | 9385.2 | 134.22 | 0.28% | 9646.0 | 30.00 | 3.07% |
| DU150-05-5 | 9506.6 | 9506.6 | 225.50 | 0.00% | 10,494.0 | 42.00 | 10.39% |
| DU150-05-6 | 9039.1 | 9039.1 | 88.17 | 0.00% | 9088.0 | 25.00 | 0.54% |
| DU150-05-7 | 9854.7 | 9869.7 | 224.73 | 0.15% | 10,017.0 | 29.00 | 1.65% |
| DU150-05-8 | 9199.3 | 9199.4 | 225.95 | 0.00% | 9550.0 | 34.00 | 3.81% |
| DU150-05-9 | 9670.9 | 9673.6 | 188.62 | 0.03% | 9972.0 | 28.00 | 3.11% |
| DU150-05-10 | 9570.6 | 9570.6 | 154.71 | 0.00% | 9794.0 | 26.00 | 2.33% |

## 5.5    Final Remarks

In this chapter, different methodologies are presented to obtain optimal or feasible solutions and lower bounds for a class of territory design problems. The class of territory design problems studied in this chapter considers different requirements: compactness, contiguity, and balanced constraints with respect to one or more

**Table 5.15** Methods to find feasible solutions for (200, 11) instances

| Instance name | Optimal value | Lagrangian relaxation | | | TDP$_{QIPR}$ | | |
|---|---|---|---|---|---|---|---|
| | | UB | CPU time | gap | UB | CPU time | gap |
| DU200-05-1 | 10,422.0 | 10,422.0 | 439.37 | 0.00% | 11,523.0 | 28.00 | 10.56% |
| DU200-05-2 | 10,639.8 | 10,648.6 | 344.58 | 0.08% | 11,425.0 | 966.00 | 7.38% |
| DU200-05-3 | 10,837.9 | 10,837.9 | 291.56 | 0.00% | 11,443.0 | 7200.00 | 5.58% |
| DU200-05-4 | 11,124.9 | 11,139.4 | 352.33 | 0.13% | 11,443.0 | 3618.00 | 2.86% |
| DU200-05-5 | 10,874.5 | 10,878.6 | 350.59 | 0.04% | 11,097.0 | 1193.00 | 2.05% |
| DU200-05-6 | 10,492.2 | 10,494.0 | 361.49 | 0.02% | 10,746.0 | 1871.00 | 2.42% |
| DU200-05-7 | 11,020.9 | 11,051.6 | 372.39 | 0.28% | 11,682.0 | 1088.00 | 6.00% |
| DU200-05-8 | 10,650.7 | 10,657.3 | 263.27 | 0.06% | 11,205.0 | 592.00 | 5.20% |
| DU200-05-9 | 11,431.3 | 11,456.9 | 302.86 | 0.22% | 11,648.0 | 1263.00 | 1.90% |
| DU200-05-10 | 11,039.5 | 11,049.2 | 353.57 | 0.09% | 11,780.0 | 2349.00 | 6.71% |

activity measures. Different models and solution procedures from the literature are presented and also new exact and approximated solution methods are proposed. The exact method is based on the formulation proposed in [19]. The exact algorithm uses a relaxation of an integer linear programming formulation of the problem, where the connectivity constraints are relaxed (because there is an exponential amount of them). This formulation is strengthened by using constraints that, although redundant, allow obtaining lower bounds of higher quality for the linear relaxation of the relaxed problem. Additionally, a cut generation procedure is utilized to obtain feasible solutions that can be used as incumbent solutions that may help reduce the enumerative effort.

Lagrangian relaxation has been extensively used in the field of location problems to find good quality dual bounds. The approximate method proposed in this chapter is based on [5] where only compactness and balance constraints were considered. In that algorithm, feasible solutions for the problem were obtained with a primal heuristic within the subgradient optimization procedure. The method proposed in this chapter also includes the connectivity constraints. Therefore, violated connectivity constraints are added to the assignment problem iteratively until a feasible assignments that also satisfy connectivity constraints is obtained.

Today, there are more efficient optimization software tools that allow the development of methodologies such as the one proposed in this chapter. These technologies lead to the development of solution methods where it is required to solve several mathematical programming subproblems in a more efficient way.

# References

1. Bergey, P.K., Ragsdale, C.T., Hoskote, M.: A simulated annealing genetic algorithm for the electrical power districting problem. Ann. Oper. Res. **121**(1–4), 33–55 (2003)
2. Bozkaya, B., Erkut, E., Laporte, G.: A tabu search heuristic and adaptive memory procedure for political districting. Eur. J. Oper. Res. **144**(1), 12–26 (2003)

3. Caro, F., Shirabe, T., Guignard, M., Weintraub, A.: School redistricting: embedding GIS tools with integer programming. J. Oper. Res. Soc. **55**(8), 836–849 (2004)
4. D'Amico, S.J., Wang, S.J., Batta, R., Rump, C.M.: A simulated annealing approach to police district design. Comput. Oper. Res. **29**(6), 667–684 (2002)
5. Díaz, J.A., Luna, D.E.: Primal and dual bounds for the vertex p-median problem with balance constraints. Ann. Oper. Res. **258**(2), 613–638 (2017)
6. Domínguez, E., Muñoz, J.: A neural model for the p-median problem. Comput. Oper. Res. **35**(2), 404–416 (2008)
7. Elizondo-Amaya, M.G., Ríos-Mercado, R.Z., Díaz, J.A.: A dual bounding scheme for a territory design problem. Comput. Oper. Res. **44**, 193–205 (2014)
8. Ferland, J.A., Guénette, G.: Decision support system for the school districting problem. Oper. Res. **38**(1), 15–21 (1990)
9. Fernández, E., Kalcsics, J., Nickel, S., Ríos-Mercado, R.Z.: A novel maximum dispersion territory design model arising in the implementation of the WEEE-directive. J. Oper. Res. Soc. **61**(3), 503–514 (2010)
10. Haase, K., Müller, S.: Upper and lower bounds for the sales force deployment problem with explicit contiguity constraints. Eur. J. Oper. Res. **237**(2), 677–689 (2014)
11. Hojati, M.: Optimal political districting. Comput. Oper. Res. **23**(12), 1147–1161 (1996)
12. Kalcsics, J.: Districting problems. In: Laporte, G., Nickel, S., Saldanha da Gamma, F. (eds.) Location Science, chap. 23, pp. 595–622. Springer, Cham (2015)
13. Kalcsics, J., Nickel, S., Schröder, M.: Towards a unified territorial design approach— Applications, algorithms and GIS integration. TOP **13**(1), 1–56 (2005)
14. Muyldermans, L., Cattrysse, D., Van Oudheusden, D., Lotan, T.: Districting for salt spreading operations. Eur. J. Oper. Res. **139**(3), 521–532 (2002)
15. Ricca, F., Simeone, B.: Local search algorithms for political districting. Eur. J. Oper. Res. **189**(3), 1409–1426 (2008)
16. Ricca, F., Scozzari, A., Simeone, B.: Political districting: from classical models to recent approaches. Ann. Oper. Res. **204**(1), 271–299 (2013)
17. Ríos-Mercado, R.Z., Fernández, E.: A reactive GRASP for a commercial territory design problem with multiple balancing requirements. Comput. Oper. Res. **36**(3), 755–776 (2009)
18. Ríos-Mercado, R.Z., López-Pérez, J.F.: Commercial territory design planning with realignment and disjoint assignment requirements. Omega **41**(3), 525–535 (2013)
19. Salazar-Aguilar, M.A., Ríos-Mercado, R.Z., Cabrera-Ríos, M.: New models for commercial territory design. Netw. Spat. Econ. **11**(3), 487–507 (2011)
20. Tavares-Pereira, F., Figueira, J.R., Mousseau, V., Roy, B.: Multiple criteria districting problems. Ann. Oper. Res. **154**(1), 69–92 (2007)

# Chapter 6
# Mathematical Programming Formulations for Practical Political Districting

**Federica Ricca and Andrea Scozzari**

**Abstract** Political districting is a very well-known technical problem related to electoral systems in which the transformation of votes into seats depends on the subdivision of the national electoral body into a given number of smaller territorial bodies. After a proper discretization of the territory, the problem consists of partitioning the territory into a prefixed number of regions which satisfy a set of geographic and demographic criteria. The problem structure falls back into one of the more general territory design problems, which arises also in other types of applications, such as school and hospital districting, sales districting, etc. In the application to political elections, the aim is to prevent districts' manipulation which may favor the electoral outcome of some specific party (*Gerrymandering*). Many political districting models and procedures have been proposed in the literature since the 1960s following different optimization strategies. Among them, many exploit mathematical programming which is one of the most used tools to solve problems in practice. The attractive feature of mathematical programming is that the model is easy-to-read, its resolution can be automated, and good compromise solutions can be computed in reasonable computational time for small and medium size problems.

## 6.1 Territory Design for Political Districting

The literature on political districting (PD) is wide and diversified, and in the last 50 years a variety of models and algorithms have been provided to face this problem through an optimization solution approach [8, 24, 25]. PD is one of the many applications of the more general territory design (TD) problem. Other applications

F. Ricca (✉)
Sapienza Università di Roma, Rome, Italy
e-mail: federica.ricca@uniroma1.it

A. Scozzari
Università degli Studi Niccolò Cusano, Rome, Italy
e-mail: andrea.scozzari@unicusano.it

© Springer Nature Switzerland AG 2020
R. Z. Ríos-Mercado (ed.), *Optimal Districting and Territory Design*, International Series in Operations Research & Management Science 284,
https://doi.org/10.1007/978-3-030-34312-5_6

can be found in a variety of real-life problems, such as the design of school and hospital districts, the regionalization of a territory for transportation service purposes, sales districting, and so on (see, [4, 12]). In all these contexts the territory is viewed as a set of small land parcels (counties, townships, wards, zip code areas, census tracts), that we call *elementary* (*territorial*) units, and each of them is characterized by a weight (referring to residents, clients, workloads). Whatever the meaning of the weights, the broad goal of TD is to obtain a set of districts as balanced as possible w.r.t. their weight (i.e., the sum of the weights of the units in the district) by collecting together contiguous territorial units. We will discuss contiguity and the other typical PD criteria in detail later. From now on we focus on territorial design for PD even if, during the presentation, references to papers dealing with the TD problem in other contexts will frequently appear.

This chapter is not meant to be a comprehensive review on mathematical programming formulations for PD, for which task previous contributions were already published [24, 25]. It was rather thought as a focus on particular modeling aspects arising in PD for the formulation of special criteria. As we will see, particular attention will be dedicated to the contiguity criterion, that plays a central role in PD, trying to give the reader an organized and systematic presentation of the main contributions in the literature on this aspect.

The PD application has attracted the interest of the researchers since the 1960s under two main aspects. On the one hand, it is always a challenge to find efficient and practical solution procedures that may be adopted to automate PD for supporting the institutional process at the basis of the political election. On the other hand, the focus is on which are the PD criteria that must be always taken into account, and which, on the contrary, should be applied only in very special cases, as happens, for example, for *communities' integrity* that may be required in countries where different ethnical/cultural communities coexist. For both types of criteria suitable modeling tools should be addressed. When implementing these criteria in a mathematical model, also the choice of an appropriate measure becomes an important issue in order to obtain a fair PD.

We study PD models assuming that the territory is divided into $n$ elementary units. In this discretized territory each unit can be identified by its (geographical) center, so that it is possible to compute distances from center to center. We also consider a graph-theoretic model for the territory that can be represented as a connected $n$-node graph $G = (N, E)$, where the nodes correspond to the elementary units and an edge between two nodes exists if and only if the two corresponding territorial units are neighboring (i.e., they share a portion of boundary). The graph $G$ is generally known as the *contiguity graph*. This was introduced by many authors independently in the 1970s (see, [2, 29]) to represent the structure of the territory, but it was frequently adopted together with a road graph that can be overlapped to $G$ to compute exact road distances instead of referring to bird's flight distances. The nodes of the contiguity graph $G$ are weighted with the population of the corresponding territorial units, while the weights of the edges represent distances between any two adjacent units.

The basic criteria for PD are four, namely *integrity, contiguity, population balance* (or *population equality*), and *compactness*. They can be stated formally as follows:

- *Integrity*—Each territorial unit cannot be split between two or more districts.
- *Contiguity*—The units of each district should be geographically contiguous, that is, one can walk from any point in the district to any other point of it without ever leaving the district.
- *Population balance* (or *population equality*)—All districts should have the same portion of representation (according to the one person-one vote principle); in particular, single-member districts should have nearly the same population.
- *Compactness*—Each district should be compact, that is, closely and neatly packed together (Oxford Dictionary). Thus, a round-shaped district is deemed to be acceptable, while an octopus- or an eel-like one is not.

The first two criteria are, in fact, strict conditions that must be satisfied in order to obtain districts that correspond to single land parcels collecting a subset of the elementary territorial units, without creating districts with holes or isolated parts.

Contiguity of the districts means that in each district one can walk for a point to another without leaving the district. It is a common geographical requirement in TD that unlikely can be ignored. Noncontiguous districts are formed by more than one portion of land separated by a land belonging to a different district. Since all TD problems are motivated by the idea of improving the management of an activity by distributing it over the territory, it is evident that non-contiguity must be avoided. Similar arguments hold for the integrity criterion that, after the discretization of the territory into $n$ population units, requires that the population of a single unit is never split between two electoral districts.

It is particularly difficult to deal with these criteria to the point that many of the models proposed in the literature do not include them formally, and the authors suggest checking integrity and contiguity a posteriori. This generally implies the need of operating manually on the district map to restore contiguity. As we will see in the following sections, a main issue in the formulation of the PD problem via mathematical programming is precisely how to model the contiguity constraints. We will discuss this aspect extensively in the second part of Sect. 6.3.

In the specific PD problem imposing contiguity is also meant as a first step towards compactness of the districts. Compactness refers to the shape of the districts, and it is evident that it is strictly related to contiguity, since hardly a district can be judged compact if it is not contiguous. Nevertheless, it must be also clear that in a district map in which contiguity holds, districts may be strongly noncompact (for example, elongated districts are contiguous but, obviously, noncompact). Compactness is a basic criterion for TD problems in general, but it is particularly important in PD for its ability to control malpractices aimed at designing biased electoral districts. It is difficult to handle, due to the wide variety of interpretations and possible measurements of compactness of a shape in general, as well as, to the difficulty in applying the geometrical notion of compactness to the articulated shapes of real territories. Including compactness (in any form) among the performance criteria of a PD model may be useful to prevent districts' manipulation, since a gerrymandered district is typically noncompact because it is the result of collecting together people voting for the same party distributed in different parts of the territory. It must be also noticed that, even if compactness may help in pursuing contiguity, it does not guarantee contiguous districts, and the role of the two criteria should be maintained separate in a PD model. Contiguity is, in fact, a structural property of the district map, and it has to be imposed as a hard constraint. On the other hand, an index of compactness typically suits the objective function, so that it can be improved as much as possible in the optimization process.

Let $n = |N|$ be the total number of territorial units and $1 < k < n$ the total number of districts, we denote by $p_i$ the population resident in territorial unit $i$, $i = 1, \ldots, n$, and by $P = \sum_{i=1}^{n} p_i$ the total population of the territory; $d_{ij}$ denotes any distance measure between unit $i$ and unit $j$. The average district population is given by $\bar{P} = \frac{P}{k}$. The PD problem can be formulated as *finding a partition of the n units into k districts according to a set of suitable criteria.*

The above four criteria are considered mandatory in the formulation of the PD problem, and they are generally included in PD models. Population equality and compactness are optimization criteria, which, in a mathematical programming approach, can be pursued in two ways, by modeling them either in the objective function, or as target constraints. This is what was actually done in the different models proposed in the literature, both in the first published papers on PD and in the more recent publications.

There are other criteria that can be also included in a PD model. One is *conformity to administrative boundaries* (or, shortly, *administrative conformity*), which prevents already existing official or normative territorial regions, which are formed by more than one elementary territorial unit (such as, counties or states), to be split between two electoral districts. It is also applicable in re-districting, when it is deemed important to maintain as much as possible the structure of the previous district map, while rebalancing the districts' population that typically changes over the years for demographic reasons. Another important (geographical) criterion is the *respect of natural boundaries*, which is useful in countries, like Italy, where the territory is characterized by many rivers, mountains, and lakes that represent obstacles for the actual reachability of the territorial units in the same district. The above two criteria are seldom considered in PD. Since it is difficult to manage many aspects in the same model, these two criteria are frequently sacrificed to privilege population equality and compactness. In fact, including all criteria in a mathematical program may result in an intractable model. In spite of this, it must be pointed out that the use of a graph-theoretic representation of the territory can help in managing this situation. Arguments similar to those already explained for guaranteeing integrity and contiguity can be addressed here. For example, the presence of a river or a lake in the common border of two territorial units $i$ and $j$ is a barrier that can be modeled directly in this graph representation. In fact, even if the territories of the two units are adjacent, in the contiguity graph it suffices to remove edge $(i, j)$ to overcome the problem. Broad discussions about political districting criteria can be found in many papers, such as in [3, 8, 12].

There are other PD criteria which are seldom considered, since they apply only in special situations and because there is no global consensus on their legitimacy. This is the case of *representation of ethnic minorities* and *respect of integrity of communities* that, in fact, can be singled out when the population of a territory is composed by different cultural or ethnic groups of citizens. The aim in this context is to maintain communities as much intact as possible in order to guarantee the right of any community present on the territory to be represented in the parliament. Differently from the classical PD criteria, there are a number of reasons that make it difficult even to formulate the mathematical constraint to model community integrity correctly. Actually, even if the principle of representation of the minorities can be accepted, it may be difficult to implement a correct interpretation of the principle in the mathematical model, and the task may be still more difficult (and less meaningful) when many different communities exist that cannot be considered the same (and, therefore, cannot be merged together in the same district). This happens, for example, in the USA, where in the same territory many communities may coexist (Hispanic, African, etc.), and integrity must be maintained for all of them without merging together any two of them. Another important issue, widely discussed in the literature, is the legitimacy of the principle itself, about supporting the representation of these communities, and how this could be done. For a more detailed discussion about this, see, for example, [8, 15, 30, 32]. In our opinion, integrity of communities can be applied when there is a single type of minority group, with a specific territorial configuration, well defined through geographic proximity of people belonging to the community. This happens, for example, in Mexico where the political representation of the Indigenous community is a problem on the Government Agenda which was also studied in many papers on PD published in the recent years (see, for example, [26]).

## 6.2  Mathematical Programming for Political Districting

The use of mathematical programming is a common practice in applications in which a quantitative approach is adopted to solve real-life problems. PD is one of these problems and the first models proposed in the literature for its solution are, in fact, mathematical programs (see Sect. 6.3). The success of mathematical programming resides in the fact that the formal model which represents the problem is easy-to-read (even if, of course, it is not as much easier to model the problem correctly). This is particularly important for PD in order to interact with politicians and lawmakers, who may feel comfortable with this kind of approach, also agreeing on the idea of solving the technical problem in a rigorous and efficient way, instead of proceeding by hands. In fact, this is what nowadays still happens in many countries, in spite of the extremely advanced and efficient technologies that are available for automated problem solving. Many criteria can be easily formulated through suitable algebraic constraints. But in real-life problems some conditions may require some additional effort to be modeled in this way, and it is possible that their correct formulation leads to a final model which is difficult to solve. This is the case of the contiguity constraints in PD, that we will discuss extensively in the following sections.

Relying on a correct and possibly concise model guarantees that in principle the problem can be solved exactly. Unfortunately, how much efficient is the solution process depends on the theoretical computational complexity of the problem. The typical situation in the general case is a computationally hard problem, but, if the size of the problem instance is not too large, the exact solution may be obtained. In the mathematics of electoral systems we already have examples. Problems related to seat allocation on the basis of the vote outcome of an election are of this type. For example, biproportional apportionment problems can be formulated as integer or mixed integer linear programs and can be solved efficiently at optimum even if the variables of the model are binary [19, 23, 25, 27]. For PD the difficulty is related to the integer decision variables and depends on the size of the territory under study. There are several cases in which the territory is small, i.e., after its discretization, the number of its territorial elementary units is small. In this case the problem can be solved easily. For medium size problems an efficient solution can be obtained using powerful computers that, thanks to the fast improvement of technology in the years, today allows solving in reasonable time problems that, for their size, were prohibitive 10 or 20 years ago. Some stratagems can be adopted to face large PD problems. One is to provide a first division of the territory into a set of subterritories, and then solve PD separately on these subterritories. This is, for example, the case of Italy, where the already existing division of the national territory into the Italian Administrative Regions is generally taken as a pre-existing subdivision which is exploited to facilitate the formation of the districts [21, 22]. In any case, due to the computational complexity of the problem, one cannot guarantee that it can be always solved to optimality, and, in some cases, a suboptimal solution must be accepted. Mathematical programming formulations help also in this case, since one possibility for solving a computationally difficult problem is to exploit the algebraic model in order to set an efficient heuristic or an approximated solution algorithm.

It must be also pointed out that, besides the traditional PD criteria, there are others which must be adopted only in some specific situations, as, for example, guaranteeing representation for cultural minorities and indigenous communities.

An additional advantage of mathematical programming is that it provides tools that may be used in a modular fashion, in the sense that the basic structure of the PD model can be set, but variants may be suggested by inserting or removing part of the constraints referring to specific requirements. As before, to avoid huge computational times, this approach can be followed with a little foresight on implications on the number of constraints and variables that are included in the final model. We will discuss this point in Sect. 6.3.2.

## 6.3    Mathematical Programming Formulations for Political Districting: Milestone Models and Recent Advances

In this section we first illustrate some papers that are considered milestones in the literature on PD, since they introduced formulations of PD which represented the starting point for many other authors. In the following years, some variants and new models were provided to be applied to real-life PD problems of different countries. We then focus on other, more recent, mathematical programming models for PD that we deem interesting especially for the different ways they suggest for tackling the contiguity constraints.

All the models refer to the formation of $k$ single-member districts for which equal population is required to guarantee fair representation. Integer and mixed integer optimization is applied, and the same kind of variables are generally used. Typically, there are two approaches to formulate PD: (1) all the $n$ elementary units can be the center of one of the $k$ districts; (2) a set of $k$ labels out of the $n$ are selected a priori, and territorial units corresponding to such labels are fixed as the centers of the $k$ districts. The two versions of the problem are generally referred to as the *noncentered* and the *centered* PD problem, respectively. In the first case, the decision variables of the models are $x_{is}, i, s = 1, \ldots, n, i \neq s$, referring to the possibility of a unit $i$ to be assigned to a center $s$ ($x_{is} = 1$) or not ($x_{is} = 0$); $x_{ss}, s = 1, \ldots, n$ is used to indicate whether unit $s$ is selected by the model as a center ($x_{ss} = 1$) or not ($x_{ss} = 0$). Following the typical notation of facility location models, some authors adopt fractional variables $0 < x_{is} \leq 1$ $i, s = 1, \ldots, n$, to measure the fraction of population of unit $i$ which is assigned to district $s$; then (binary) indicator variables $y_s$ $s = 1, \ldots, n$ are considered for deciding whether unit $s$ is selected as a center ($y_s = 1$) or not ($y_s = 0$). When the center units are fixed in advance, the set of units $N$ is partitioned a priori into two subsets, one for the centers, and another for units that are not centers. We denote them by $S$ and $U$, respectively, so that we have $S \cup U = N, S \cap U = \emptyset$, with $|S| = k, |U| = n - k$.

### 6.3.1    Classical PD Formulations

The study of PD through mathematical programming can be dated back to the early 1960s, when Weaver and Hess published their first work on this topic [31]. In a second paper, Hess et al. presented their PD model and solution approach more formally, and published it in an operations research (OR) journal [10]. It is worth noting that, in the same years, Hess and Samuels published another paper on territory design but applied to sales districts [9]. The approach suggested in these papers basically proposes to adapt techniques for warehouse location-allocation problems to PD. The years that followed were characterized by a rich production of OR papers on PD.

The paper by Hess et al. [10] is generally considered as the earliest paper applying OR, and, in particular, mathematical programming techniques, to PD. Let $n$ be the total number of territorial units and $k$ the number of districts, the problem is formalized as a discrete location problem. The idea is to identify $k$ units representing the centers of the $k$ districts, so that each territorial unit must be assigned to exactly one district center. The model has $n^2$ binary variables, $x_{is}, i, s = 1, \ldots, n$: for $i \neq s$, $x_{is}$ is equal to 1 when unit $i$ is assigned to center $s$, and 0 otherwise; the variable $x_{ss}$ is equal to 1 whenever unit $s$ is chosen as one of the centers. The political districting model is the following:

$$\min \quad \sum_{i=1}^{n} \sum_{s=1}^{n} d_{is}^2 \, p_i \, x_{is} \tag{6.1a}$$

$$\text{subject to} \quad \sum_{s=1}^{n} x_{is} = 1 \qquad\qquad i = 1, \ldots, n \tag{6.1b}$$

$$\sum_{s=1}^{n} x_{ss} = k \tag{6.1c}$$

$$a \, \bar{P} \, x_{ss} \leq \sum_{i=1}^{n} p_i \, x_{is} \leq b \, \bar{P} \, x_{ss} \qquad s = 1, \ldots, n \tag{6.1d}$$

$$x_{is} \in \{0, 1\}, \qquad\qquad i, s = 1, \ldots, n \tag{6.1e}$$

where $p_i$ is the population of unit $i$, $d_{is}$ is the distance between unit $i$ and center $s$, and $a$ and $b$ are the minimum and the maximum allowable district population fractions, calculated as a percentage of the average district population $\bar{P}$. The first $n$ constraints (6.1b) mean that each unit must belong exactly to one district. The next one (6.1c) imposes that the total number of districts is exactly $k$. The last group of $2n$ constraints (6.1d) refers to the upper and lower bounds on district population imposed to control population balance. The objective function (total inertia) is a measure of compactness.

Due to the computational complexity of the above problem, a solution is found heuristically by applying an iterative procedure that, in fact, relies on a different model, i.e., a transportation model with continuous variables. At each iteration the procedure performs the following steps: (1) makes a guess for the $k$ district centers; (2) solves a transportation problem to form districts with population exactly equal to $\bar{P}$ and assigns units to such districts at minimum cost (the assignment cost of unit $i$ to district centered in $s$ is given by $d_{is}^2$); (3) adjusts the solution of the transportation problem by assigning any split unit $i$ entirely to the district to which the largest quota of its population was already assigned; (4) computes the centroids of the current districts to update the district centers to be used in the next iteration. Steps (1)–(4) are repeated until the centers do not change in two successive iterations.

The above solution approach has two main drawbacks. The first one is that solving the transportation problem at step 2 generally implies that units are split between two or more districts. In fact, step 3 is included to restore integrity of the districts. The second problem is that the contiguity of the districts is neither modeled in the integer program (6.1a)–(6.1e), nor it is considered in the heuristic solution procedure. This obviously requires an a posteriori revision for spatial contiguity. In addition, we observe that the theoretical convergence of the procedure is not guaranteed.

In spite of the above problems, [10] has always been considered as a seminal paper in mathematical programming applied to PD, and it has traced the starting point of a long and wide literature production on this subject. An example is the paper by Hojati [11] who, starting from [10], developed an alternative approach based on a capacitated facility location model. The author suggests a three-phase procedure, in which Phase 2 corresponds to the one given in [10]. Actually, instead of adopting an iterative strategy based on successive adjustments, the centers are located only once at the beginning of the procedure and this choice is permanent. This, in fact, corresponds to changing from a noncentered to a centered version of the problem. The author introduces a mixed integer warehouse location model aimed at locating the $k$ district centers on the basis of the Euclidean distances $d_{is}$. The model is the following:

$$\min \quad \sum_{i=1}^{n}\sum_{s=1}^{n} d_{is}^2\, p_i\, x_{is} \tag{6.2a}$$

$$\text{subject to} \quad \sum_{s=1}^{n} x_{is} = 1 \qquad i = 1,\ldots,n \tag{6.2b}$$

$$\sum_{i=1}^{n} p_i\, x_{is} = \bar{P}\, y_s \qquad s = 1,\ldots,n \tag{6.2c}$$

$$\sum_{s=1}^{n} y_s = k \tag{6.2d}$$

$$x_{is} \le y_s \qquad i, s = 1,\ldots,n \tag{6.2e}$$

$$0 \le x_{is} \le 1 \qquad i, s = 1,\ldots,n \tag{6.2f}$$

$$y_s \in \{0, 1\} \qquad s = 1,\ldots,n \tag{6.2g}$$

A Lagrange relaxation of the model is derived and it is solved by a subgradient optimization algorithm. The solution of the above program provides the $k$ district's centers to be used in Phase 2. When, after the solution of the transportation problem, there are split territorial units (i.e., units fractionally assigned to more than one center), the author introduces the *split resolution problem* (SRP) which is formulated as a graph-theoretic model. Actually, he takes into consideration the subgraph of the transportation graph whose vertices are given by the split units on one side, and by those centers to which some split units have been (partially) assigned, on the other side. The author shows that SRP is NP-hard by a reduction from the partition problem [6], and suggests an heuristic procedure to solve SRP (Phase 3 of his procedure) based on the solution of a sequence of capacitated transportation problems defined over a suitable modified network (see, for details [11]). Notice that, applying this heuristic does not guarantee integrity of territorial units but only reduces as much as possible the number of split ones. It must be also pointed out that, as in [10], also Hojati does not consider the contiguity criterion, so that noncontiguous district maps may arise in the final solution, requiring additional effort to fix this problem.

In 1970, Garfinkel and Nemhauser also proposed a two-phase procedure which models and solves PD via a set partitioning approach [7]. Phase 1 generates the set $J$ of feasible districts that are actually considered in Phase 2 when formulating the set partitioning model. The set $J$ includes only contiguous districts that satisfy some requirements related to population balance and compactness. A value $\alpha \in [0, 1]$ is fixed as the maximum tolerance on the absolute deviation of each district population from $\bar{P}$, and district $s$ is feasible only if $|P_s - \bar{P}| \le \alpha \bar{P}$. The same tolerance $\alpha$ is used again in Phase 2 in the formulation of the objective function of the set partitioning model. The districts included in $J$ are also feasible w.r.t. compactness, according to an index based on both the maximum distance between two territorial units in the district and the district area, whose value should be less than or equal to a fixed threshold $\beta$. Compactness is taken into account only in the first phase.

Phase 1 applies an implicit enumeration strategy to find all the feasible solutions to include in $J$. A contiguity graph $G$ is adopted to represent the territory and the district generation is performed via a tree-search on $G$. Starting from an arbitrary node (unit), subtrees (which correspond to districts) are formed by including in the district under construction nodes adjacent to some other nodes already in the district. This is performed until the district population becomes feasible w.r.t. the threshold value $\alpha \bar{P}$. Checks are made periodically to avoid formation of enclaves, that is, groups of units that remain isolated after the construction of the current district.

In Phase 2 the following model is formulated to select from $J$ a set of $k$ districts that covers each population unit exactly once and minimizes the overall deviation of district populations from $\bar{P}$:

$$\min \quad \sum_{s \in J} f_s x_s \tag{6.3a}$$

$$\text{subject to} \quad \sum_{s \in J} a_{is} x_s = 1 \qquad i = 1, \dots, n \tag{6.3b}$$

$$\sum_{s \in J} x_s = k \tag{6.3c}$$

$$x_s \in \{0, 1\} \qquad s \in J \tag{6.3d}$$

where $x_s = 1$ if district $s \in J$ is chosen and $x_s = 0$ otherwise. The $a_{is}, i = 1, \ldots, n,$ $s = 1, \ldots, |J|$ are the typical coefficients of a set covering or partitioning problem where $a_{is} = 1$ if unit $i$ is in district $s$ and $a_{is} = 0$ otherwise.

Coefficient $f_s$ is given by the absolute value of the ratio between the deviation of the population of district $s$, $P_s$, from the average district population $\bar{P}$ and a fraction $\alpha$ of $\bar{P}$, $\alpha \in [0, 1]$, i.e., $f_s = (|P_s - \bar{P}|)/(\alpha \bar{P})$, and it is introduced to optimize population balance in the objective function of the model. Due to the computational complexity of the set partitioning problem, implicit enumeration is applied also for solving the problem of Phase 2.

Other authors followed the approach in [7], such as Merhotra et al. [14] who still use a set partitioning approach, but adopt a different objective function in order to take into account also compactness of the overall district map.

## 6.3.2   Recent and New Ideas for PD Modeling

In this section we consider only the four basic criteria for PD and refer to a PD model based on a contiguity graph $G$. Different models have been proposed in the recent literature on PD according to the specific way one chooses to manage population balance and compactness, but in all cases, relying on a connected graph $G$, guarantees that integral and contiguous districts can be found by searching for a connected partition of $G$. In fact, the weakness of PD models illustrated in Sect. 6.3.1 is that they do not consider contiguity explicitly. On the contrary, in this section we will see how contiguity can be handled in the model through appropriate algebraic constraints. As already discussed by Duke et al. for TD [5], there are three main approaches for modeling contiguity in a mathematical program: (i) tree-based; (ii) order-based; (iii) flow-based. All approaches were, in fact, followed by many authors who studied TD in general, but interesting formulations were provided in particular for PD. In (i) the feasibility problem is formalized as finding a spanning forest of $G$, and each district corresponds to a subtree in the forest. The model requires cycle preventing constraints analogous to those adopted in the mathematical formulation of the minimum spanning tree problem of a graph, or those used to avoid subtours in the traveling salesmen problem. The tree-based models are therefore characterized by a big number of constraints which grows rapidly as the number $n$ of territorial units increases. This implies that this kind of formulation cannot be usefully adopted in practice even for small size problem instances ($n \leq 50$, see [5]). For this reason, in the present paper we do not discuss this approach, but focus on the other two. According to (ii), contiguity of a district is guaranteed by following a specified order to include a unit in a district: such order implies that unit $i$ can be assigned to a district centered in $s$ either if $i$ is adjacent to $s$, or if there exists a unit $j$ adjacent to $i$ already assigned to the same district. The big number of *order constraints* which arise in this case can be reduced according to different strategic rules that simplify the problem at the beginning. Adopting these rules implies that the final solution is not guaranteed to be optimal w.r.t. the original model, but remains meaningful for PD and can be obtained in reasonable computational times for up to medium size problems. Finally, in (iii) the existence of the underlying contiguity graph has naturally suggested the idea of exploiting network flow constraints for modeling contiguity in a PD mathematical program. In this case, the graph $G$ must be preliminary directed in the ordinary way obtaining a network $\mathcal{G} = (N, A)$ in which an edge of $G$ connecting nodes $i$ and $j$ is replaced in $\mathcal{G}$ by a pair of arcs $(i, j)$ and $(j, i)$. The rationale behind the model is to consider a different flow for each district to be formed, with the condition that any two different flows cannot use the same arc.

In the following, we focus our attention on the two approaches (ii) and (iii) and discuss different mathematical programming models for contiguous PD. As we will see, different model variants can be considered to solve optimally PD problems of small size, while additional strategies and techniques can be addressed to efficiently obtain contiguous district maps also for larger size instances.

### 6.3.2.1 Flow-Based Contiguous PD Models

In [28] Shirabe models the topological property of a contiguous district via a set of flow constraints which are able to enforce the property independently of other possible constraints that may be included in the mathematical program. The set of contiguity constraints is formulated for a single district, but can be replicated $k$ times in the model when there are $k > 1$ districts to be formed. Starting from this structural set of constraints, a PD model can be obtained by suitably modeling also population equality and compactness. In the following, we first illustrate how the set of contiguity constraints is formulated in [28] for a single district, and then we report on some specific PD models suggested in the same work which attracted our interest.

The set of constraints to form a single contiguous district proposed by Shirabe takes into account the condition that a maximum allowable number $m$ of territorial units can be included in the district. In the network flow approach each node of the network $\mathcal{G}$ has a supply of one unit of flow. Then, the district is identified as a subgraph of $\mathcal{G}$ with a specific node serving as a sink, which receives one unit of flow from every other node of $\mathcal{G}$ that is included in the district. The demand of the sink node is bounded by a parameter $m$. A first set of variables is given by $x_i, i = 1, \ldots, n$ which is equal to 1 if units $i$ are included in the district under formation, and 0 otherwise; variables $w_i, i = 1, \ldots, n$ are also included in the model to decide whether unit $i$ is the sink of the district ($w_i = 1$) or not ($w_i = 0$). In addition, flow variables $y_{ij} \geq 0$ indicate the amount of flow on arc $(i, j) \in A$. The set of flow constraints proposed in [28] for the formation of a single contiguous district is the following:

$$\sum_{j|(i,j)\in A} y_{ij} - \sum_{j|(j,i)\in A} y_{ji} \geq x_i - m w_i \qquad i \in N \qquad (6.4a)$$

$$\sum_{j|(j,i)\in A} y_{ji} \leq (m-1)x_i \qquad\qquad i \in N \qquad\qquad (6.4b)$$

$$\sum_{i\in N} w_i = 1 \qquad\qquad (6.4c)$$

$$x_i \in \{0, 1\} \qquad\qquad i \in N \qquad\qquad (6.4d)$$

$$w_i \in \{0, 1\} \qquad\qquad i \in N \qquad\qquad (6.4e)$$

$$y_{ij} \geq 0 \qquad\qquad (i, j) \in A \qquad\qquad (6.4f)$$

The first group of constraints (6.4a) controls the net inflow at node $i$, either when $i$ is the sink, or when it is simply a unit which was included in the district under formation. When $i$ is the sink of the district, we have $w_i = x_i = 1$ and the constraint becomes:

$$\sum_{j|(j,i)\in A} y_{ji} - \sum_{j|(i,j)\in A} y_{ij} \leq m - 1$$

Since the outflow $\sum_{j|(i,j)\in A} y_{ij}$ of a sink $i$ is 0, the constraint actually limits the inflow of the sink by a maximum of $m - 1$ and, in this case, this constraint is the same as the corresponding one for the sink $i$ in the second group (6.4b). The second group of constraints, in fact, bounds above by $(m - 1)$ the incoming flow of *any* node $i$ included in the district ($x_i = 1$). When $i$ is in the district, but it is not the sink node ($x_i = 1$, $w_i = 0$), the constraint of the first group requires that the net outflow of $i$ is at least one (i.e., the unit supply of $i$). If $i$ is not in the district ($x_i = 0$), the flow constraints altogether impose that no flow passes through $i$. Notice that the second group of flow constraints also models the condition that flow cannot reach node $i$ if unit $i$ is not included in the district (dependency of variables $y_{ji}$ from $x_i$). Then constraint (6.4c) guarantees that only one node in the district plays the role of the sink. The remaining constraints specify the nature of the model variables.

We note that in the above formulation any node in $N$ may play the role of the sink to which the flow is directed. Therefore, it must be pointed out that in this model the sink is not necessarily a central node of the district. In spite of this, when in PD the centers of the districts are suitably fixed before districting, they can be adopted as the sinks of the model. For a given center (sink) $s$, the previous set of constraints becomes:

$$\sum_{j|(i,j)\in A} y_{ij} - \sum_{j|(j,i)\in A} y_{ji} = x_i \qquad i \in N, i \neq s \qquad (6.5a)$$

$$\sum_{j|(j,i)\in A} y_{ji} \leq (m-2)x_i \qquad i \in N, i \neq s \qquad (6.5b)$$

$$\sum_{j|(j,s)\in A} y_{js} \leq (m-1) \tag{6.5c}$$

$$x_i \in \{0, 1\} \qquad\qquad i \in N, i \neq s \tag{6.5d}$$

$$x_s = 1 \tag{6.5e}$$

$$y_{ij} \geq 0 \qquad\qquad (i, j) \in A, i \neq s \tag{6.5f}$$

$$y_{sj} = 0 \qquad\qquad j \in N, (s, j) \in A \tag{6.5g}$$

Since the sink node is known, the flow constraints (6.5a) are specialized for $i \neq s$, and for the units that belong to the district they impose that the net outflow in $i$ is *exactly* equal to 1.

When the PD problem is centered, i.e., there is a set $S$ of units fixed in advance as the centers of the $k$ districts, the above set of constraints (6.5a)–(6.5g) can be embedded in a mixed integer linear program (MILP) that also takes into account the other PD criteria. Considering compactness in the objective function, and bounding the population of each district from below and above, [28] provides the following model where the set of decision variables extends to $x_{is}, i = 1, \ldots, n, s = 1, \ldots, k$, which are equal to 1 if unit $i$ is assigned to district $s$ and 0 otherwise. In this case the model becomes a multi-commodity flow in which there are $k$ different types of flow, one for each district. Then, also the flow variables are extended to $y_{ijs}$ which is the amount of flow related to district $s$ passing through arc $(i, j) \in A$. The model is the following:

$$\min \sum_{s\in S}\sum_{i\in N} d_{is}x_{is} \tag{6.6a}$$

$$\text{subject to} \quad \sum_{j|(i,j)\in A} y_{ijs} - \sum_{j|(j,i)\in A} y_{jis} = x_{is} \qquad s \in S, i \in U \tag{6.6b}$$

$$\sum_{j|(j,i)\in A} y_{jis} \leq (m_s - 2)x_{is} \qquad s \in S, i \in U \tag{6.6c}$$

$$\sum_{j|(j,s)\in A} y_{jss} \leq (m_s - 1) \qquad s \in S \tag{6.6d}$$

$$\sum_{i\in N} p_i x_{is} \geq (1 - \alpha)\bar{P} \qquad s \in S \tag{6.6e}$$

$$\sum_{i\in N} p_i x_{is} \leq (1 + \alpha)\bar{P} \qquad s \in S \tag{6.6f}$$

$$x_{is} \in \{0, 1\} \qquad s \in S, i \in U \tag{6.6g}$$

$$x_{ss} = 1 \qquad s \in S \tag{6.6h}$$

$$x_{s's} = 0 \qquad s, s' \in S, s \neq s' \tag{6.6i}$$

$$y_{ijs} \geq 0 \qquad\qquad s \in S, \, i, j \in U \qquad (6.6j)$$

$$y_{iss} \geq 0 \qquad\qquad s \in S, \, i \in U \qquad (6.6k)$$

where $d_{is}$ is the distance between unit $i$ and center $s$, while $\alpha$ is the maximum allowable absolute deviation from the average district population. The flow constraints are exactly the same as in (6.5a)–(6.5g), but here they are replicated $k$ times, one for each center $s \in S$. The population balance constraints (6.6e) and (6.6f) follow, together with some conditions on assignment variables that guarantee that a center unit $s$ is always assigned to itself, and it is never assigned to a district centered in a $s' \neq s$.

For the sake of simplicity, in the above model we did not report an additional set of constraints on variables $y_{ijs}$ which are considered in [28]. Such constraints impose that the flow from a center $s'$ to another center $s$ adjacent to $s'$ is equal to 0. In a PD model, they can be discarded if one guarantees that, during the selection of the district centers, any two centers are never located on adjacent nodes of $G$. This is a common rule in PD since the selection of the centers is generally based on a maximum dispersion criterion (see, for example, [22]).

Variants of the model can be obtained by swapping the role of compactness and population equality in the model, or formulating compactness in a different way (for details, see [28]).

We observe that in model (6.6a)–(6.6k) population balance constraints are combined with bounds on the maximum number of flow units which may reach the sink of a district ($m_s$). In real life applications it must be payed attention to the fact that this may lead to infeasible models, depending on the values of the input parameters $m_s$, $s \in S$, the actual population of the units, $p_i$, $i = 1, \ldots, n$, and the value of $\alpha$. To illustrate this point, consider the simple example in which $n = 10$ and $k = 2$, with $\bar{P} = 10$ and populations of the 10 units given by $9, 2, 1, 1, 1, 1, 1, 1, 1, 1$. Let $m_1 = 3$, $m_2 = 7$, with $\alpha = 0.05$. In this case, to satisfy population balance, the unit with population equal to 9 would be forced to stay alone in a district. Therefore, the second district would include all the other units (9 units). This is obviously a nonfeasible solution, since the model allows including in a district at most $m_2 = 7$ units. This simple example shows that, even if it is true that the contiguity flow constraints proposed in [28] can be embedded in any MILP formulation of PD, a prudent choice of the above parameters must be performed to guarantee model feasibility.

We also observe that another possible variant of model (6.6a)–(6.6k) can be obtained for the case in which the number of territorial units to be included in a district is not limited above. This can be done by setting $m_s = n - k + 1$ for all $s \in S$. Notice that, in this case, feasibility problems discussed above do not hold any more.

In [5] different integer and mixed integer linear programs for the general TD problem with exactly $p$ districts (called *p-regions problem*) are provided, with a focus on the different possibilities for modeling contiguity. In particular, one of them proposes a variant of the contiguity constraints provided in [28]. They still consist of a set of flow constraints, but with a slight difference in how the flow is controlled over the network. The model does not take into account population balance, and considers compactness in the objective function which is given by the sum of the distances between all pairs of units belonging to the same district. For the sake of comparison, in the model formulation that follows we adopt the same notation as in the previous illustrated models. We denote the total number of districts by $k$ instead of $p$ that is used in [5]. We also continue to denote by $x_{is}$ the assignment variables to decide whether unit $i$ is assigned to district centered in $s$ or not, and by $y_{ijs}$ the flow related to district $s$ passing through arc $(i, j)$. The new variables $t_{ij}$ introduced by Duque et al. [5] to model the objective function are binary and are set to 1 if units $i$ and $j$ belong to the same district, and 0 otherwise. This makes it possible to model a linear objective function in the variables $t_{ij}$.

The mathematical program is the following:

$$\min \quad \sum_{i \in N} \sum_{j \in N: j > i} d_{ij} t_{ij} \tag{6.7a}$$

$$\text{subject to} \quad \sum_{s \in S} x_{is} = 1 \qquad\qquad i \in N \tag{6.7b}$$

$$w_{is} \le x_{is} \qquad\qquad s \in S, i \in N \tag{6.7c}$$

$$\sum_{i \in N} w_{is} = 1 \qquad\qquad s \in S \tag{6.7d}$$

$$\sum_{j|(i,j) \in A} y_{ijs} - \sum_{j|(j,i) \in A} y_{jis}$$

$$\ge x_{is} - (n - k) w_{is} \qquad s \in S, i \in N \tag{6.7e}$$

$$y_{ijs} \le x_{is}(n - k) \qquad\qquad s \in S, i \in N, j|(i, j) \in A \tag{6.7f}$$

$$y_{ijs} \le x_{js}(n - k) \qquad\qquad s \in S, i \in N, j|(i, j) \in A \tag{6.7g}$$

$$t_{ij} \ge x_{is} + x_{js} - 1 \qquad\qquad s \in S, i, j \in N, j > i \tag{6.7h}$$

$$x_{is} \in \{0, 1\} \qquad\qquad s \in S, i \in N \tag{6.7i}$$

$$w_{is} \in \{0, 1\} \qquad\qquad s \in S, i \in N \tag{6.7j}$$

$$t_{ij} \ge 0 \qquad\qquad i, j \in N, j > i \tag{6.7k}$$

$$y_{ijs} \ge 0 \qquad\qquad s \in S, i \in N, j|(i, j) \in A \tag{6.7l}$$

Variables $t_{ij}$ are, in fact, integer, but this has not to be imposed explicitly in the model, since it is implied by the set of constraints (6.7f)–(6.7h) and by the minimization of the objective function.

In this model, the sink nodes are not fixed in advance, so that variables $w_{is}$ ($i = 1, \ldots, n; s = 1, \ldots, k$) are necessary to model the choice of the sinks. Since $k$ districts must be formed, the implication $w_{is} \le x_{is}$ for any pair $i$ and $s$ must be included in the model to guarantee that a unit that is chosen as the sink of district $s$ will be necessarily assigned to $s$. The flow constraints are similar to those in [28], but not exactly the same, since in the first group (6.7e) $m_s$ is replaced by $(n - k + 1)$, for all $s \in S$. In fact, this corresponds to not bounding the maximum number of territorial units in a district at all, leaving the possibility that all units but $k - 1$ are included in one single district (as happened in our previous example). The remaining $k - 1$ would be the centers of the remaining districts. The other two groups of constraints (6.7f) and (6.7g) involving the flow variables are a different way to model the dependency of $y_{ijs}$ from $x_{is}$. Thus, in the model, the flow on the arcs of $G$ is not bounded above, but the presence of these constraints, and of those related to the dependency between $t_{ij}$ and $x_{is}$ and $x_{js}$, together with the minimization of the objective function, guarantees that only the strictly necessary quantity of flow passes through the arcs.

The fact that no population balance constraints are included makes the model slimmer than, for example, (6.6a)–(6.6k). This implies that the computational times reported in [5] are underestimated if one wants to understand how much time is needed to solve optimally a real-life PD problem by using a general-purpose optimization software to solve this model.

### 6.3.2.2  Order-Based PD Contiguous Models

As already discussed before in this section, contiguity constraints can be modeled in different ways. In the following we illustrate how to formulate contiguity via order constraints for a PD problem in which the centers of the districts are fixed in advance. We start from the models proposed in [1] and [13] where the authors apply mathematical programming to the problem of partitioning a tree $T = (N, E)$ into $k$ connected components, each including exactly one center. Since the underlying graph is a tree, contiguity of the districts can be explicitly formulated by a polynomial number of order constraints. The idea is that in a tree contiguity can be accomplished by imposing that if unit $i$ is included in the district centered in $s$, then all units in $T$ lying in the unique path $P_{i,s}$ from $i$ to $s$ must be included in the same district as $i$. The model has $O(n^2 k)$ order constraints in total. This can be further improved to $O(n\,k)$ if one imposes order constraints on successive adjacent nodes in $P_{i,s}$ and exploits transitivity, thus obtaining the following set of constraints:

$$x_{is} \leq x_{j(i,s),s} \qquad i \in U,\ s \in S,\ (i,s) \notin E \qquad (6.8)$$

$$x_{is} \in \{0, 1\} \qquad i \in U,\ s \in S. \qquad (6.9)$$

where the binary variables $x_{is}$ have the same meaning as before, and $j(i, s)$ is the label of the node $j$ adjacent to $i$ in the unique path from $i$ to $s$. In the above papers the objective function is a linear function of the assignment variables, with coefficients given by a cost function $c : U \times S \to \Re$ which associates a cost $c_{is}$ to each pair $(i, s)$, $i \in U,\ s \in S$. These are, in fact, *flat* costs, that is, costs independent of the topology of the underlying graph. When the same model is formulated with a cost metric function the problem can be solved in strongly polynomial time even if the input graph is arbitrary (for details, see [1]). Additional theoretical results on the computational complexity of this problem are provided in [1], showing that the problem becomes intractable when the cost function is not a metric. One result states that, when $c_{ij}$ are flat costs, the problem is NP-complete even if $c$ is monotone and the input graph is bipartite. A second result is related to the capacitated version of the problem in which a weight is associated to each node of the graph, and the sum of the weights of the nodes in each class of the partition is bounded above and below. In [13] it is shown that this problem is NP-complete on 2-spider trees even if $c$ is a metric cost function.

The above results become relevant for PD applications when the model objective function is based on flat costs, or when population balance is imposed by lower and upper bounds on the district populations; in the latter case the PD model corresponds to the capacitated version of the partitioning problem studied in [1] and [13]. From the above discussion a positive result is that, when the costs are metric and no capacity (population balance) constraints are included in the model, the problem is easy to solve on general graphs. Both in [1] and in [33] it is, in fact, stated that an optimal solution can be obtained in this case simply by assigning each territorial unit to its closest center. Unfortunately, in PD applications, where the graph $G$ representing the territory is typically a planar graph, population balance is a main criteria which must be included in the model.

We now present a PD formulation which exploits order constraints for modeling contiguity and includes population balance. The objective function is a measure of compactness based on distances $d_{is}, i \in U, s \in S$.

Consider the contiguity graph $G = (N, E)$, with $N = U \cup S$. For a given $s \in S$, let $T_s$ be the tree rooted at $s$ and formed by the minimum cardinality paths from any $i \in U$ to $s$; denote by $E_s$ the set of its edges. We consider a tree $T_s$ for each $s \in S$. Basing on these trees, for every unit $i$ and center $s$ we can compute univocally the distance $d_{is}$ as the cardinality of the (unique) path form $i$ to $s$ in $T_s$. The PD problem is then formulated as follows:

$$\min \quad \sum_{s \in S} \sum_{i \in U} p_j \, d_{is}^2 \, x_{is} \tag{6.10a}$$

$$\text{subject to} \quad \sum_{i \in N} p_i x_{is} \geq (1 - \alpha) \bar{P} \qquad s \in S \tag{6.10b}$$

$$\sum_{i \in N} p_i x_{is} \leq (1 + \alpha) \bar{P} \qquad s \in S \tag{6.10c}$$

$$\sum_{s \in S} x_{is} = 1 \qquad i \in U \tag{6.10d}$$

$$x_{is} \leq x_{j(i,s),s} \qquad i \in U, s \in S, (i, s) \notin E_s \tag{6.10e}$$

$$x_{is} \in \{0, 1\} \qquad i \in U, s \in S \tag{6.10f}$$

In spite of its computational difficulty, the advantage of this formulation is that it takes into account all the basic PD criteria. In addition, using the trees $T_s$, $s \in S$, contiguity can be formulated by introducing in the model only a polynomial number of order constraints ($O(nk)$, the same number needed when $G$ is a tree). The hard constraints in the model are the ones related to population balance (6.10b) and (6.10c). The solution approach is then to perform a Lagrange relaxation of these constraints. The resulting model has a flat cost objective function and includes only assignment and order constraints. It is then possible to show that an equivalent formulation of the relaxed model as a vertex packing with continuous variables can be obtained. A straightforward solution of a continuous vertex packing can be found by bringing the model back to a network flow one, and then solving it through techniques proposed in [16, 17]. Here we do not report all the details of these models, which, in fact, are the subject of a work in progress finalized to their application to real-life PD problems. The aim is to understand how much the computational time to solve practical PD can be reduced via this kind of approach, and which levels of population balance and compactness of the districts can be reached [20].

To conclude this section, we point out that a similar approach was proposed by Zoltners and Sinha some years ago for the sales territory alignment problem [33]. Even if their model arises in a different application context, it is applicable also to PD, since all the basic PD criteria are taken into account. The model by Zoltners and Sinha fits the PD application when there is just one type of attribute associated to each territorial unit and it is specific of this unit and independent of the districts. In the model, the attribute of unit $i$ is measured by a weight $a_i$, $i = 1, \ldots, n$ that may correspond to different types of weigh associated to $i$ (the number of residents, of clients, etc.). In the application to PD we have $a_i = p_i$. Even if more than one theoretical model is presented in [33], results are provided in particular for the one which adopts equality constraints for population balance, that is, the population of each district must be exactly equal to $\bar{P}$. In the following we report this model using the same notation adopted in the previous models:

$$\min \quad \sum_{i=1}^{n} \sum_{s=1}^{k} d_{is} \, p_i \, x_{is} \tag{6.11a}$$

$$\text{subject to} \quad \sum_{i=1}^{n} p_i \, x_{is} = \bar{P} \qquad s = 1, \ldots, k \tag{6.11b}$$

$$\sum_{s=1}^{k} x_{is} = 1 \qquad i = 1, \ldots, n \tag{6.11c}$$

$$x_{is} \leq \sum_{h \in A_{is}} x_{hs} \qquad i = 1, \ldots, n, \quad s = 1, \ldots, k \tag{6.11d}$$

$$x_{is} \in \{0, 1\} \qquad i = 1, \ldots, n, \quad s = 1, \ldots, k. \tag{6.11e}$$

In the above model $d_{is}$ are road distances. They are used in order to take into account geographic features of the territory that may produce obstacles for the reachability between two units that are adjacent in the contiguity graph $G$.

Comparing this model with (6.10a)–(6.10f), two main differences arise. First, the population balance constraint is managed through strict equations in (6.11a)–(6.11e), while it is formulated by upper and lower bounds in (6.10a)–(6.10f). Second, in (6.11a)–(6.11e) the notion of *hierarchical adjacency tree* is introduced instead of the trees $T_s$, $s \in S$ adopted in (6.10a)–(6.10f). A hierarchical adjacency tree rooted at $s$ is obtained starting from the shortest path tree rooted at $s$ (computed basing on the road distances), but including additional edges chosen from those belonging to near optimal shortest paths from some $i$ to $s$. This implies that some nodes are duplicated on different branches of the tree, and that a node $i$ can have more than one parent in the hierarchical adjacency tree. In the model, $A_{is}$ denotes the set of nodes which immediately precede node $i$ on any branch of the hierarchical adjacency tree. The model allows including a unit $i$ in a district centered in $s$ in different ways, that is, forcing different sets of other units lying on the same path from $i$ to $s$ to be included in the same district. If, on the one hand, this avoids the rigid contiguity constraints used in (6.10a)–(6.10f), in which unit $i$ can be connected to center $s$ only using the shortest path, on the other hand, the choice of how many and which additional edges have to be included in the hierarchical tree is arbitrary. If in sales territory alignment this choice can be legitimately made by the firm, in PD application a neutral rule should be established, since the structure of the hierarchical tree influences the district formation, and this clearly may produce consequences on the final electoral outcome. For this reason, even if we deem model (6.11a)–(6.11e) interesting and elegant under a theoretical viewpoint, we believe it cannot be applied for PD in practice.

In both models (6.10a)–(6.10f) and (6.11a)–(6.11e), the difficult constraints are those related to population balance, either when they are equalities or inequalities, and, in fact, the solution approach is based on their Lagrange relaxation, so that the resulting relaxed model maintains only the easy constraints, i.e., assignment and order constraints, that can be managed easier than in the original model. In particular, in (6.11a)–(6.11e), relying on metric distances, the optimal solution of the relaxed model can be obtained in a straightforward way exploiting the theoretical results recalled at the beginning of this section. The Lagrange relaxation of (6.10a)–(6.10f) is a bit more complicated, but it can be still shown that, after this relaxation and by successive reformulations, the model can be re-written as a computationally easy-to-solve equivalent mathematical program. This model, that, at the moment, is under study and has to be still validated through empirical experimentation, seems to be promising for PD applications [20].

## 6.4 Conclusions and New Perspectives in Political Districting Modeling

In this chapter we discussed how mathematical programming can be usefully applied to solve the particular territory design problem related to political districting, and in which way mathematical programs help in dealing with the impartiality purposes arising in a so much delicate application. It must be noticed that, even if the practical problem is to divide a national territory into smaller territories according to typical geographical and population-based criteria, in the context of political elections any of these aspects acquire a particular importance, since, differently from the commercial territory design problem, only unbiased criteria should be pursued, the aim being to guarantee a fair division of the territory which, combined with the specific electoral system, should ensure to the competing parties total neutrality of the vote translation into seats. This is very difficult to achieve and, in fact, neutrality cannot be guaranteed by any mathematical tool, nor it can be shown theoretically that a district map is free from any bias or distortion in this sense. But what a mathematical model can surely do is to measure criteria objectively, and, once an agreement on a set of criteria is reached, it is guaranteed that using a mathematical program is a good way to implement them and to reach the best value at least for one via the optimization of the model's objective function. It is also clear that the support that a mathematical model can provide is most of all in the possibility of solving in short times very large combinatorial problems, so that, even if an intervention by the human judge and experience is always necessary (and recommended), a fast and correct solution procedure allows starting from a high quality district plan, and easily producing alternative ones, if necessary. This is probably the point that motivates the constant lively production of mathematical programming models for PD in the literature, especially for specific PD problems arising in some countries. In fact, there are many countries, like Mexico, that recognized that PD is a technical problem that can be solved in a rigorous and efficient way, instead of by hands, and consequently decided to adopt an automated PD procedure. In other countries, automation is used but without exploiting the power of the mathematical modeling with much effort and a clear waste of time.

We recall that the size of the problem may undermine the efficiency of the solution process, but in many countries the problem size is small, so that adopting a mathematical program solves the problem exactly. Another possibility to tackle together computational times and correctness of the solution is to split the PD problem of the national territory into a set of smaller PD problems on smaller subterritories, such as, for example, regions, or counties. This multiplies the number of PD problems to be solved, but drastically decreases the computational time for solving each problem, and may be a viable tool to obtain exact optimal solutions in short times.

To conclude, it is natural to ask ourselves which may be the possible future developments in this field. From an analysis of the recent literature, we observed a slowdown in the proposal of innovative ways for modeling PD criteria. Managing the efficient solution of the problem seems to be the main challenge at the moment. To improve this point, the wide variety of works focuses on applying heuristic procedures, such as local search [3, 21], or artificial bee colony based algorithms [26], that are able to find contiguous district maps with a good compromise between population balance and compactness in short times. This is motivated by the fact that the solution time is a critical factor in PD applications, and it increases fast as the number $n$ of territorial units increases, but the difficulty of the model also depends on the high number of constraints that the model must include when contiguity is explicitly formalized together with the other basic PD criteria. In view of this, a possible approach to improve the efficiency of the model is to apply a row generation strategy in which difficult constraints are ignored at the beginning and added later (and gradually) only when a solution is obtained that does not satisfy some of them. This approach is advisable for models in which contiguity is formalized through order constraints, and exactly these generally are the constraints to be ignored at the beginning. This approach basically follows a *relaxation by elimination* strategy. Another possibility to make the model tractable is relaxing some constraints according to Lagrange. For example, it is well-known that including in the PD model bounds on the district population is difficult to manage, and, in fact, in many cases a Lagrange relaxation of these constraints is performed to ease the model solution process (Sect. 6.3.2).

In spite of the above considerations, we believe that mathematical programming modeling itself can be further exploited for PD in different ways. In this presentation we observed that models based on network flow constraints for contiguity are generally easier to handle than those in which this condition is formalized through tree-based or order constraints. Flow constraints for contiguity can be modeled in different ways, and some are more efficient than others (see, Sect. 6.3.2). Therefore, one possible future development could be introducing new models for contiguous districting where contiguity constraints are still realized through flow constraints, but different from (and, hopefully, more efficient than) the already existing ones. We ourselves are working on this subject, providing and testing new mathematical programming formulations for centered and noncentered partition problems of general graphs that can be also applied to PD. Experimentation is in progress on these new models.

# References

1. Apollonio, N., Lari, L., Puerto, J., Ricca, F., Simeone, B.: Polynomial algorithms for partitioning a tree into single-center subtrees to minimize flat service costs. Networks **51**(1), 78–89 (2008)
2. Bodin, L.D.: A district experiment with a clustering algorithm. Ann. N. Y. Acad. Sci. **219**(1), 209–214 (1973)
3. Bozkaya, B., Erkut, E., Laporte, G.: A tabu search heuristic and adaptive memory procedure for political districting. Eur. J. Oper. Res. **144**(1), 12–26 (2003)

4. Duque, J.C., Ramos, R., Suriñach, J.: Supervised regionalization methods: a survey. Int. Reg. Sci. Rev. **30**(3), 195–220 (2007)
5. Duque, J.C., Church, R.L., Middleton, R.S.: The $p$-regions problem. Geogr. Anal. **43**(1), 104–126 (2011)
6. Garey, M.R., Johnson, D.S.: Computers and intractability. In: A Guide to the Theory of NP-Completeness. Freeman, New York (1979)
7. Garfinkel, R.S., Nemhauser, G.L.: Optimal political districting by implicit enumeration techniques. Manag. Sci. **16**(8), 495–508 (1970)
8. Grilli di Cortona, P., Manzi, C., Pennisi, A., Ricca, F., Simeone, B.: Evaluation and optimization of electoral systems. In: SIAM Monographs on Discrete Mathematics and Applications. SIAM, Philadelphia (1999)
9. Hess, S.W., Samuels, S.A.: Experiences with a sales districting model: criteria and implementation. Manag. Sci. **18**(4), 41–54 (1971)
10. Hess, S.W., Weaver, J.B., Siegfelatt, H.J., Whelan, J.N., Zitlau, P.A.: Nonpartisan political redistricting by computer. Oper. Res. **13**(6), 998–1006 (1965)
11. Hojati, M.: Optimal political districting. Comput. Oper. Res. **23**(12), 1147–1161 (1996)
12. Kalcsics, J., Nickel, S., Schröder, M.: Towards a unified territorial design approach—applications, algorithms and GIS integration. Top **13**(1), 1–74 (2005)
13. Lari, I., Ricca, F., Puerto, J., Scozzari, A.: Partitioning a graph into connected components with fixed centers and optimizing cost-based objective functions or equipartition criteria. Networks **67**(1), 69–81 (2016)
14. Mehrotra, A., Johnson, E.L., Nemhauser, G.L.: An optimization based heuristic for political districting. Manag. Sci. **44**(8), 1100–1114 (1998)
15. Morril, R.L.: Making redistricting models more flexible and realistic. Oper. Geogr. **9**(1), 2–9 (1991)
16. Nemhauser, G.L., Trotter, L.E. Jr.: Properties of vertex packing independence system polyhedra. Math. Program. **6**(1), 48–61 (1974)
17. Nemhauser, G.L., Trotter, L.E. Jr.: Vertex packings: structural properties and algorithms. Math. Program. **8**(1), 232–248 (1975)
18. Nygreen, B.: European assembly constituencies for wales comparing of methods for solving a political districting problem. Math. Program. **42**(1–3), 159–169 (1988)
19. Pukelsheim, F., Ricca, F., Scozzari, A., Serafini, P., Simeone, B.: Network flow methods for electoral systems. Networks **59**(1), 73–88 (2012)
20. Ricca, F., Scozzari, A.: Ordered-based formulations for political districting. Working Paper. Università degli Studi di Roma, La Sapienza, Rome (2018)
21. Ricca, F., Simeone, B.: Local search algorithms for political districting. Eur. J. Oper. Res. **189**(3), 1409–1426 (2008)
22. Ricca, F., Scozzari, A., Simeone, B.: Weighted Voronoi region algorithms for political districting. Math. Comput. Model. **48**(9–10), 1468–1477 (2008)
23. Ricca, F., Scozzari, A., Serafini, P., Simeone, B.: Error minimization methods in biproportional apportionment. Top **20**(3), 547–577 (2012)
24. Ricca, F., Scozzari, A., Simeone, B.: Political districting: from classical models to recent approaches. Ann. Oper. Res. **204**(1), 271–299 (2013)
25. Ricca, F., Scozzari, A., Serafini, P.: A guided tour of the mathematics of seat allocation and political districting. In: Endriss, U. (ed.) Trends in Computational Social Choice, chap. 3, pp. 49–68. ILLC, University of Amsterdam, Amsterdam (2017)
26. Rincón García, E.A., Gutiérrez Andrade, M.A., de-los-Cobos-Silva, S.G., Ponsich, A., Mora-Gutiérrez, R.A., Lara-Velázquez, P.: A system for political districting in the state of Mexico. In: Sidorov, G., Galicia-Haro, S. (eds.) Advances in Artificial Intelligence and Soft Computing. Lecture Notes in Computer Science, vol. 9413, pp. 248–259. Springer, Cham (2015)
27. Serafini, P., Simeone, B.: Parametric maximum flow methods for minimax approximation of target quotas in biproportional apportionment. Networks **59**(2), 191–208 (2012)
28. Shirabe, T.: Districting modeling with exact contiguity constraints. Environ. Plann. B. Plann. Des. **36**(6), 1053–1066 (2009)

29. Simeone, B.: Optimal graph partitioning. In: Atti delle Giornate di lavoro AIRO (Associazione Italiana di Ricerca Operativa), pp. 57–73. Urbino (1978)
30. Thoreson, J.D., Liittschwager, I.M.: Computers in behavioral science. Legislative districting by computer simulation. Behav. Sci. **12**(3), 237–247 (1967)
31. Weaver, J.B., Hess, S.W.: A procedure for nonpartisan districting: development of computer techniques. Yale Law J. **73**(1), 288–308 (1963)
32. Williams, J.C. Jr.: Political redistricting: a review. Pap. Reg. Sci. **74**(1), 13–40 (1995)
33. Zoltners, A.A., Sinha, P.: Sales territory alignment: a review and a model. Manag. Sci. **29**(11), 1237–1257 (1983)

# Chapter 7
# Multi-Period Service Territory Design

**Matthias Bender and Jörg Kalcsics**

**Abstract** In service districting, a given set of customers has to be assigned to the individual members of the service workforce such that each customer has a unique representative, each service provider faces an equitable workload and travel time, and service districts are compact and contiguous. One important, but rarely addressed feature of many service districting applications is that customers require service with different frequencies. As a result, planners not only have to design the service districts, but also schedule visits to customers within the planning horizon such that the workload for each service provider is the same across all periods and the set of all customers visited in the same time period is as compact as possible.

We present a mixed-integer linear programming formulation for the problem. As it turns out, only very small data sets can be solved to optimality within a reasonable amount of time. One of the reasons for that appears to be the high level of symmetry between solutions. We first characterize these symmetries and propose ideas to try to eliminate them in the formulation. Afterwards, we focus on the scheduling component of the problem and present a location-allocation based heuristic for determining visiting schedules for the service providers for fixed districts. In addition, we propose a branch-and-price algorithm to solve larger data sets to proven optimality. One of the novel features of the algorithm is a symmetry-reduced branching scheme that results in a significant speed-up.

M. Bender
Department of Logistics and Supply Chain Optimization, Research Center for Information Technology (FZI), Karlsruhe, Germany
e-mail: mbender@fzi.de

J. Kalcsics (✉)
School of Mathematics, The University of Edinburgh, Edinburgh, UK
e-mail: joerg.kalcsics@ed.ac.uk

© Springer Nature Switzerland AG 2020 129
R. Z. Ríos-Mercado (ed.), *Optimal Districting and Territory Design*, International Series in Operations Research & Management Science 284,
https://doi.org/10.1007/978-3-030-34312-5_7

## 7.1 Introduction

In service districting, a given set of customers, each with a known on-site service time, has to be assigned to the individual members of the service workforce such that each customer has a unique representative and each service provider faces an equitable workload and travel time. One of the most important applications of service districting is sales territory design. According to [21], approximately every tenth full-time employee in the USA is working as a field and retail sales person and the expenditure for them is more than three trillion dollars every year. Territories with a low sales potential, intense competition, or too many small customers lead to low morale, poor performance, a high turnover rate, and an inability to assess the productivity of individual territories. Therefore, well-planned decisions are required to enable an efficient market penetration and decrease costs and improve customer service and sales. A good territory design can increase sales by an estimated 2–7% compared to an average solution [21]. Other applications of service districting include determining districts for technical maintenance [5], home visits by health-care personnel [3, 4], postal or leaflet delivery [6, 7], and waste collection, salt spreading, and winter gritting [9, 16, 18], just to name a few. For more examples, see [12].

The three main planning criteria in service districting are balance, contiguity, and compactness. Balance describes the desire for districts that incur an equal workload for each member of the service workforce. An uneven distribution of workload among service providers will often result in discontent and, subsequently, a decrease in productivity and service quality provided to customers. In most models, balance is treated as a hard constraint and measured in terms of the deviation of the actual district workload from the average workload across all territories. Contiguous districts are desired to obtain clearly defined geographic areas of responsibility for each service provider. This is especially important in sales territory design, as it is not uncommon for sales persons to compete for customers with a high sales potential. Compactness describes the desire for districts that are geographically closely packed. Apart from the visual appeal of compact districts, the criterion often serves, together with contiguity, as a proxy for minimizing travel times. The hope is that compact and contiguous districts result on average in smaller travel times on a day-to-day basis than non-compact and/or non-contiguous districts. While the motivation for the latter two criteria is very intuitive, it is very hard to rigorously define and assess them. Customers are typically represented as points on a map, making it difficult to determine whether a district is contiguous or not. Within heuristic methods, a common approach is to deem a district as being contiguous, if no customer served by another provider lies within the convex hull of the set of all customers of the district. While being visually appealing, it is very difficult to include this measure in mathematical programming formulations. The alternative is to derive proximity graphs that allow for an assessment of contiguity in a graph-theoretic sense. See [12, 19, 20] for detailed discussions. Even more ambivalence and ambiguity can be found for compactness. Given that its main motivation is to propagate short daily travel distances, an obvious measure would be to calculate the actual travel distances, i.e., tours, within the district. This is, however, usually not sensible from a computational as well as practical point of view. Concerning the former, real-world data sets for districting problems are often quite large, easily comprising several thousand customers. This renders the actual calculation of tours very time-consuming, especially in the presence of additional requirements, like time windows or multi-day trips. With respect to the latter, service territory design is a tactical planning problem, with districts usually being re-designed at most once per year. According to estimates, about one in five visits needs to be rescheduled to another day due to short-term requests of customers or unexpected events [1]. This renders the benefit of calculating daily or even weekly routes questionable. As a result, most districting models assess compactness using geometric measures, like the Schwartzberg or Reock test, or distance-based measures, e.g., the sum of pairwise distances between all customers of the district. See [10] for a recent review of compactness measures. Finally, we note that as non-contiguous districts are typically less compact than contiguous ones, contiguity is oftentimes not explicitly modelled as a criterion. Instead, the hope is that just ensuring compactness suffices to obtain also contiguous districts.

One important, but rarely addressed feature of many service districting applications is that customers require service with different frequencies. Some customers have to be visited weekly, while others require service only once per month. As a result, planners not only have to design the service districts, but also schedule visits to customers within the planning horizon. For example, if the planning horizon is divided into weeks and days, then we also have to decide which customers should be visited in which week and on which day of that week. The criteria for scheduling customer visits are very similar to the ones for designing districts. Concerning balance, the total workload incurred by all customers served in each time period should be the same across all periods. Moreover, the set of all customers visited in the same time period should be as compact as possible to reduce travel times during each period. While contiguity is still desirable, differing visiting frequencies will make it very difficult, or even impossible, to obtain contiguous sets of customers for each period.

The remainder of the chapter is organized as follows. In the next section, we formally introduce the service territory design problem and briefly review related literature. In Sect. 7.3 we present a mathematical formulation for the problem and discuss practically important variants. In Sect. 7.4 we focus on the scheduling component of the problem and sketch a heuristic and an exact algorithm for solving the problem. We end the chapter with some conclusions and an outlook to future work.

## 7.2 Problem Description and Literature Review

We consider two sets of time periods: *weeks* and *days*. Let $W = \{1, \ldots, |W|\}$ and $D = \{1, \ldots, |D|\}$ denote the set of weeks and days, respectively, in the planning horizon, and $D^w \subseteq D$ the set of days in week $w \in W$. To simplify the exposition, we assume that each week contains the same number of days, $|D^w| = m \in \mathbb{N}$, $w \in W$. Note that $D$ consecutively numbers all days of the planning horizon, i.e., $|D| = m \cdot |W|$. The set of *customers* is denoted by $B = \{1, \ldots, |B|\}$ and each customer $b \in B$ requires on-site service by a service provider once every $r_b \in \mathbb{N}$ weeks. If customer $b$ is served in week $w$, then we call this a *visiting week* or *service week* for $b$, analogously for days. $r_b$ is called the *week rhythm* of customer $b \in B$. Each customer must be served within the first $r_b$ weeks and then regularly every $r_b$ weeks. For example, if $|W| = 8$ and $r_b = 2$, then there exist two feasible patterns for the visiting weeks of $b$, namely $(1, 3, 5, 7)$ and $(2, 4, 6, 8)$. For a customer with a week rhythm of $r_b = 3$, the feasible patterns are $(1, 4, 7)$, $(2, 5, 8)$, and $(3, 6)$. While the first customer will be served exactly four times during the planning horizon, the second will be served either twice or thrice, depending on the first visiting week. To avoid this ambiguity, we assume that $|W|$ is the least common multiple of all week rhythms. Consequently, customer $b$ is served exactly $|W|/r_b$ times during the planning horizon. We denote by $s_b \in \mathbb{R}^+$ the *service time* per visit for customer $b \in B$. The set of *districts* or *territories* is defined as $T = \{1, \ldots, |T|\}$ and each district is patronized by a single service provider. We assume that each district has a designated center, e.g., the office or home of the service provider. For $t \in T$, we denote by $B_t$ the set of customers assigned to territory $t$, by $B_t^w \subseteq B_t$ the set of customers of territory $t$ visited in week $w$, and by $B_t^d \subseteq B_t^w$ the set of customers of territory $t$ visited on day $d \in D^w$, $w \in W$. We call the latter two *week clusters* and *day clusters* of district $t$. The problem can now be stated as follows:

The **Multi-Period Service Territory Design Problem (MPSTDP)** seeks to partition the set of customers into service territories (*districting subproblem*) and to determine the visiting weeks and days for each customer (*scheduling subproblem*) such that all districts, week clusters, and day clusters are balanced and compact.

Balanced week and day clusters are desirable to avoid deviations from the regular working hours for the service workforce. The motivation for compact day clusters is to allow for short day-to-day travel times of the service providers and give them flexibility to sequence visits according to (short-term) customer requests, e.g., because of time windows. Due to unexpected events, service visits often have to be rescheduled to another day. Having compact week clusters makes it easier for a service provider to flexibly re-schedule missed visits on a given day to one of the remaining days of the week without having to make overly long detours. While the districts may only be re-designed every 2 or 3 years, the scheduling problem is often solved more frequently in order to adapt to shifts in the customer base during a year. For example, some customers have dropped out, others have been newly acquired by the service provider, and yet others request more (or less) frequent visits. If the territories are already given and only the scheduling subproblem has to be solved, we denote the problem as MPSTDP-S.

The MPSTDP was first introduced in Bender et al. [1] and was motivated by a project with the company PTV Group,[1] a commercial provider of geomarketing software, among other tools. After introducing the MPSTDP, the authors focus on the scheduling subproblem MPSTDP-S, although in a slightly more general form: service times can depend on the visiting weeks and customers may request to be served more than once per week. They introduce a mixed-integer linear programming formulation for the MPSTDP-S and present a location-allocation heuristic to solve the problem for large instances. Using real-world data sets, they show that the solutions provided by the heuristic are in most cases significantly better than the solutions of the—then current—heuristic of the company, both in terms of compactness and balance. In a subsequent paper, Bender et al. [2] present a branch-and-price algorithm for the MPSTDP-S that includes efficient techniques to reduce the inherent symmetries in the scheduling subproblem. They show that the branch-and-price algorithm is able to solve real-world data sets by two orders of magnitude faster than Gurobi's branch-and-bound method.

A related problem is discussed in Lei et al. [14]. The authors assume that customer demand is dynamic over time, i.e., changes from period to period, where each time period may comprise several weeks. Each customer has to be served once per period. For each time period, the task is to partition all customers into districts, and then subsequently partition the customers of each district into day clusters. The quality of solutions is measured based on the total number of districts, the compactness of day clusters, the similarity of districts across time periods, and the equity of salesmen profits. To solve the problem, the authors combine the four measures into a single objective function and then present an adaptive large neighborhood search algorithm. In a follow-up paper, Lei et al. [15] generalize the problem by allowing customer demand to be either deterministic or stochastic, where the actual demand realization for stochastic customers is only known after the districts have been planned. Instead of aggregating the four assessment measures for solutions, they propose a multi-objective evolutionary algorithm to determine non-dominated solutions. Mourgaya and Vanderbeck [17] discuss a variant of the periodic vehicle routing problem. While customers still have to be allocated to vehicles, computing the actual vehicle tours is no longer part of the problem. Instead, for each time period, they aim to obtain compact sets of customers for each vehicle and to balance workload between vehicles. To solve the problem, they propose a branch-and-price algorithm, in which the pricing problems are based

---

[1] www.ptvgroup.com.

on quadratic knapsack problems and solved using a greedy heuristic. For a more detailed review, including also less closely related problems and applications, we refer to Bender et al. [1, 2].

## 7.3 Mathematical Formulations

In this section we present a mixed-integer linear programming formulation for the MPSTDP that extends the ones in [1, 2] by the districting subproblem. We start with the general MPSTDP. Afterwards, we point out several deficiencies inherent to the problem that make it difficult to solve it optimally. Finally, we consider the case that customers also require regularity with respect to the visiting days and not just the visiting weeks.

### 7.3.1 A Mathematical Formulation for the MPSTDP

Before we start presenting the formulations, we have to introduce a few more notations. For $i, b \in B$ and $t \in T$, let $c_{bi}$ and $\bar{c}_{bt}$ be the travel distance (or time or cost) between customer $i$ and $b$ and between customer $b$ and the designated center of district $t$, respectively. To measure the compactness of a set $B' \subseteq B$ of customers, we use the classical approach of calculating the sum of distances of all customers in $B'$ to a *center* of $B'$. On a district level, this center is the designated district center. Unfortunately, this center cannot be used for measuring weekly and daily compactness, as each configuration of week and day clusters would give exactly the same value. Thus, to measure the compactness of week and day clusters in a meaningful way, we determine an artificial center for each. Typically, this artificial center is the customer that minimizes the sum of distances to all customers in $B'$: $i^* = \arg\min_{i \in B} \sum_{b \in B'} c_{bi}$. Note that the artificial center does not necessarily have to belong to $B'$. We call the corresponding customer the *week center* or *day center* of week $w$ or day $d$, respectively. Moreover, we denote by $LB_t$ and $UB_t$ the lower and upper bound, respectively, of the acceptable total service time for the service provider of district $t$. Analogously, we define $LB_t^w$, $UB_t^w$, and $LB_t^d$, $UB_t^d$ as the acceptable service time in week $w$ and on day $d$, respectively. Finally, let $fv(b, w) = ((w - 1) \bmod r_b) + 1$; if customer $b$ is served in week $w$, then $fv(b, w) \in \{1, \ldots, r_b\}$ is the week of the first visit to $b$.

To formulate the problem, we define the following sets of decision variables:

$$g_{bt}^w = \begin{cases} 1, & \text{if customer } b \text{ is part of district } t \text{ and served in week } w \\ 0, & \text{otherwise.} \end{cases}$$

$$h_{bt}^d = \begin{cases} 1, & \text{if customer } b \text{ is part of district } t \text{ and served on day } d \\ 0, & \text{otherwise.} \end{cases}$$

$$x_{it}^w = \begin{cases} 1, & \text{if customer } i \text{ is the week center of district } t \text{ in week } w \\ 0, & \text{otherwise.} \end{cases}$$

$$y_{it}^d = \begin{cases} 1, & \text{if customer } i \text{ is the day center of district } t \text{ on day } d \in D \\ 0, & \text{otherwise.} \end{cases}$$

$$u_{bit}^w = \begin{cases} 1, & \text{if customer } b \text{ is assigned to week center } i \text{ of district } t \text{ in week } w \\ 0, & \text{otherwise.} \end{cases}$$

$$v_{bit}^d = \begin{cases} 1, & \text{if customer } b \text{ is assigned to day center } i \text{ of district } t \text{ on day } d \\ 0, & \text{otherwise.} \end{cases}$$

A mixed-integer linear programming formulation for the MPSTDP is then given as (MPSTDP)

$$\min \sum_{t \in T} \sum_{b \in B} \left( \alpha_1 \sum_{w \in W} \bar{c}_{bt} g_{bt}^w + \alpha_2 \sum_{i \in B} \sum_{w \in W} c_{bi} u_{bit}^w + \alpha_3 \sum_{i \in B} \sum_{d \in D} c_{bi} v_{bit}^d \right) \tag{7.1a}$$

$$\text{subject to} \quad \sum_{t \in T} \sum_{w=1}^{r_b} g_{bt}^w = 1 \qquad\qquad b \in B \tag{7.1b}$$

$$\sum_{b \in B} \sum_{w \in W} s_b g_{bt}^{fv(b,w)} \leq UB_t \qquad\qquad t \in T \tag{7.1c}$$

$$\sum_{b \in B} \sum_{w \in W} s_b g_{bt}^{fv(b,w)} \geq LB_t \qquad\qquad t \in T \tag{7.1d}$$

$$\sum_{i \in B} u_{bit}^w = g_{bt}^{fv(b,w)} \qquad\qquad b \in B,\, w \in W,\, t \in T \tag{7.1e}$$

$$u_{bit}^w \leq x_{it}^w \qquad\qquad i, b \in B,\, w \in W,\, t \in T \tag{7.1f}$$

$$\sum_{i \in B} x_{it}^w = 1 \qquad\qquad w \in W,\, t \in T \tag{7.1g}$$

$$\sum_{b \in B} s_b g_{bt}^{fv(b,w)} \leq UB_t^w \qquad\qquad w \in W,\, t \in T \tag{7.1h}$$

$$\sum_{b \in B} s_b g_{bt}^{fv(b,w)} \geq LB_t^w \qquad\qquad w \in W,\, t \in T \tag{7.1i}$$

$$\sum_{d \in D^w} h_{bt}^d = g_{bt}^{fv(b,w)} \qquad\qquad b \in B,\, w \in W,\, t \in T \tag{7.1j}$$

$$\sum_{i \in B} v_{bit}^d = h_{bt}^d \qquad\qquad b \in B,\, d \in D,\, t \in T \qquad\qquad (7.1\mathrm{k})$$

$$v_{bit}^d \le y_{it}^d \qquad\qquad i, b \in B,\, d \in D,\, t \in T \qquad\qquad (7.1\mathrm{l})$$

$$\sum_{i \in B} y_{it}^d = 1 \qquad\qquad d \in D,\, t \in T \qquad\qquad (7.1\mathrm{m})$$

$$\sum_{b \in B} s_b h_{bt}^d \le U B_t^d \qquad\qquad d \in D,\, t \in T \qquad\qquad (7.1\mathrm{n})$$

$$\sum_{b \in B} s_b h_{bt}^d \ge L B_t^d \qquad\qquad d \in D,\, t \in T \qquad\qquad (7.1\mathrm{o})$$

$$g_{bt}^w,\, h_{bt}^w \in \{0, 1\} \qquad\qquad b \in B,\, d \in D,\, w \in W,\, t \in T \qquad (7.1\mathrm{p})$$

$$x_{it}^w,\, y_{it}^d \in \{0, 1\} \qquad\qquad i \in B,\, d \in D,\, w \in W,\, t \in T \qquad (7.1\mathrm{q})$$

$$u_{bit}^w,\, v_{bit}^d \ge 0 \qquad\qquad b, i \in B,\, d \in D,\, w \in W,\, t \in T \qquad (7.1\mathrm{r})$$

The objective function (7.1a) represents a weighted average of the compactness of districts, week clusters, and day clusters, where $\alpha_1, \alpha_2, \alpha_3 \ge 0$ are weighting factors. By summing over all weeks when computing district compactness, we put a higher emphasis on customers with small week rhythms, who will be included in proportionally more week and day clusters than customers with a large value for $r_b$. Constraints (7.1b) ensure that each customer $b$ must be assigned to exactly one district and served exactly once within the first $r_b$ weeks. Note that due to the rigid week rhythms, it suffices to determine the first visiting week for each customer. The next two sets of constraints, (7.1c) and (7.1d), guarantee that the total workload per district is within the given bounds. Constraints (7.1e) and (7.1f) enforce that a customer is allocated to the corresponding week center in each visiting week. Constraints (7.1g) select a week center for each week and district. Constraints (7.1h) and (7.1i) ensure that the weekly workload is within the specified limits. Constraints (7.1j) make sure that there is a visiting day in each visiting week of a customer. Constraints (7.1k)–(7.1o) are the daily analogues of Constraints (7.1e)–(7.1i). The last three constraints specify the variable domains.

If the districts are already given and we just want to solve the scheduling subproblem MPSTDP-S, then we obtain a formulation for the latter by simply removing all references to district $t$ in formulation (MPSTDP), including the index $t$ in all parameters and variables, as well as Constraints (7.1c) and (7.1d).

## 7.3.2 Computational Experiments

To evaluate the computational tractability of the formulation, we present some preliminary experiment results in the following for three different variants of the problem on randomly generated data sets. The first variant is the MPSTDP without days, i.e., only the visiting weeks but not the visiting days have to be scheduled (the main reason for omitting days is that we are struggling to generate feasible data sets for small numbers of customers). The second variant is the scheduling subproblem, also without days. And the last is the scheduling subproblem with days, i.e., the MPSTDP-S. We generate data sets with 30 and 40 customers. The customer and service provider locations are uniformly distributed over the squares $[0, 10]^2$ and $[2, 8]^2$, respectively. The service times are uniformly drawn from the interval $[40, 80]$ and the travel distances $c_{bi}$ and $\bar{c}_{bt}$ are based on the Euclidean distance between the respective points. We use six different patterns of week rhythms with differing planning horizons. The patterns are shown in Table 7.1. The number of weekdays is five, except for the last two types, where we have to go down to 3 days as we are again struggling to generate feasible data sets for the MPSTDP-S. For each pattern and number of customers, we generate five data sets.

For the two scheduling subproblems, we set $LB^w = 0.8\mu^{week}$ and $UB^w = 1.2\mu^{week}$ as lower and upper bound, respectively, for the feasible range of weekly service times, where $\mu^{week} = \frac{1}{|W|} \sum_{b \in B} s_b \frac{|W|}{r_b} = \sum_{b \in B} \frac{s_b}{r_b}$ is the average service time per week (recall that $|W|/r_b$ is the total number of visits to customer $b \in B$). Analogously, we define $LB^d = 0.6\mu^{day}$ and $UB^d = 1.4\mu^{day}$, with $\mu^{day} = \frac{1}{|D|} \sum_{b \in B} s_b \frac{|W|}{r_b}$ for days. We use the same allowed weekly deviation for the MPSTDP without days, but with $\hat{\mu}^{week} = \frac{1}{|T|} \sum_{b \in B} \frac{s_b}{r_b}$. Concerning the allowed deviation from the average district size $\mu^{district} = \frac{1}{|T|} \sum_{b \in B} s_b \frac{|W|}{r_b}$, we had to set $LB^t = 0$ and $UB^t = 2\mu^{district}$ to avoid infeasibility of the instances. Concerning the weighting factors in the objective function, for the two scheduling subproblems we use $\alpha_2 = \frac{1}{3}$ and $\alpha_3 = \frac{2}{3}$, putting a higher emphasis on daily compactness. For the MPSTDP, we use $\alpha_1 = \frac{1}{3}$ and $\alpha_2 = \frac{2}{3}$. All tests are carried out on a Windows 7 machine with an Intel i5-6500T processor and 8 GB memory. The formulations are solved using CPLEX 12.7.1 with a time limit of 30 min.

We present the results in Table 7.2. The first column indicates the week rhythm pattern. The next three columns present the results for formulation (MPSTDP)

| Table 7.1 Overview of the six different patterns of week rhythms | Type | Week rhythms | No. weeks | No. weekdays |
|---|---|---|---|---|
| | 1 | $\{1, 2, 4\}$ | 4 | 5 |
| | 2 | $\{2, 4\}$ | 4 | 5 |
| | 3 | $\{1, 2, 4, 8\}$ | 8 | 5 |
| | 4 | $\{2, 4, 8\}$ | 8 | 5 |
| | 5 | $\{1, 2, 4, 8, 16\}$ | 16 | 3 |
| | 6 | $\{2, 4, 8, 16\}$ | 16 | 3 |

**Table 7.2** Experiment result for the three variants of the problem

| | MPSTDP w/o days | | | MPSTDP-S w/o days | | | MPSTDP-S | | |
|---|---|---|---|---|---|---|---|---|---|
| Pattern | Gap | Opt | Time | Gap | Opt | Time | Gap | Opt | Time |
| *30 customers* | | | | | | | | | |
| 1 | 0.00% | 5 | 11.8 | 0.00% | 5 | 0.6 | 0.01% | 5 | 291.1 |
| 2 | 1.23% | 4 | 738.3 | 0.00% | 5 | 1.3 | 3.94% | 1 | 1492.8 |
| 3 | 0.00% | 5 | 274.5 | 0.01% | 5 | 2.1 | 5.84% | 0 | 1800.7 |
| 4 | 0.01% | 5 | 556.8 | 0.00% | 5 | 5.2 | 17.06% | 0 | 1800.0 |
| 5 | 2.63% | 3 | 1194.1 | 0.00% | 5 | 6.2 | 11.05% | 0 | 1800.1 |
| 6 | 6.87% | 3 | 1265.7 | 0.00% | 5 | 14.9 | 12.83% | 0 | 1800.1 |
| **Summary** | **1.76%** | **25** | **655.6** | **0.00%** | **30** | **5.1** | **8.45%** | **6** | **1497.5** |
| *40 customers* | | | | | | | | | |
| 1 | 0.01% | 5 | 16.5 | 0.00% | 5 | 0.8 | | | |
| 2 | 0.01% | 5 | 12.0 | 0.00% | 5 | 3.1 | | | |
| 3 | 0.01% | 5 | 435.7 | 0.00% | 5 | 4.7 | | | |
| 4 | 0.01% | 5 | 1056.8 | 0.00% | 5 | 13.8 | | | |
| 5 | 0.00% | 5 | 1538.1 | 0.01% | 5 | 16.2 | | | |
| 6 | 0.00% | 5 | 1201.6 | 0.00% | 5 | 74.2 | | | |
| **Summary** | **0.01%** | **30** | **710.1** | **0.00%** | **30** | **18.8** | | | |

without days. The columns labelled $Gap$, $Opt$, and $Time$ give the MIP gap reported by CPLEX upon termination, the number of instances with a proven optimal solution, and the run time in seconds, respectively. The time and gap values are averages over the five instances. The next two triples of columns show the results for the MPSTDP-S without days and the MPSTDP-S. We group the results by number of customers and summarize them for each number, showing the average MIP gap and runtime and the total number of instances solved to proven optimality.

Starting with the latter, we can see that only the first set of instances with pattern 1 can all be solved optimally within half an hour for the 30 customer instances. For all other patterns, CPLEX hits the runtime limit of 30 min without finding a proven optimal solution for all but one instance. Thus, we refrain from carrying out tests for 40 customers. Turning to the MPSTDP-S without days, the data sets are much easier to solve. Now, all instances can be solved optimally within seconds. While a reduction in running times compared to MPSTDP-S was to be expected, the level of reduction is slightly surprising. Including the districting subproblem, the running times increase considerably, as can be seen from the results for the MPSTDP without days. While most of the instances can still be solved optimally within 30 min, the increase in difficulty is apparent.

Looking at the impact that the length of the planning horizon has on the results, not unexpectedly the run times increase as $|W|$ increases. A common observation across all results is that the data sets for patterns 2, 4, and 6 are—in most cases— harder to solve than their counterparts from patterns 1, 3, and 5, respectively. A possible explanation is that the odd numbered patterns contain weekly customers, whereas the even numbered patterns do not. For weekly customers, there is no scheduling decision to be made on a weekly level, only on a daily level, reducing the number of decision variables in the formulation.

## 7.3.3 Symmetry

As we have just seen, already small data sets with just 30 customers are difficult to solve optimally within half an hour. One of the reasons for this intractability appears to be the high level of symmetry that is inherent in the scheduling subproblem [1, 2]. In the following, we will analyze this more formally, focussing on the scheduling subproblem MPSTDP-S only, i.e., all references to territories are omitted. We say that two solutions $(B^w)_{w \in W}$ and $(\hat{B}^w)_{w \in W}$ are *week-symmetric*, if there exists a permutation $\pi : W \to W$ such that $B^{\pi(w)} = \hat{B}^w$, $w \in W$ (recall $B^w$ and $\hat{B}^w$ denote the respective week clusters). Analogously, we say that 2 week clusters $B^w = (B^d)_{d \in D^w}$ and $\hat{B}^w = (\hat{B}^d)_{d \in D^w}$ are *day-symmetric with respect to week* $w$, if there exists a permutation $\sigma : D^w \to D^w$ such that $B^d = \hat{B}^{\sigma(d)}$, $d \in D^w$ (recall that $D^w$ is the set of days in week $w$). Two solutions are called *day-symmetric* if they are day-symmetric with respect to each week.

Starting with days, any permutation of days within the same week $w$ results in a day-symmetric solution with respect to $w$ with exactly the same objective value. Hence, a week cluster $B^w$ incurs $m! - 1$ day-symmetric solutions with respect to $w$ (recall that $m = |D^w|$). For example, for 5 weekdays we obtain $5! - 1 = 119$ alternate solutions. Considering all weeks of the planning horizon, each solution has $(m!)^{|W|} - 1$ day-symmetric solutions. Even for just a 4-week horizon with 5 days per week, this results in more than 207 million day-symmetric solutions.

In addition, there is also symmetry on the level of weeks, albeit much less. For the ease of exposition, we assume in the following that all week rhythms are powers of two, i.e., $r_b = 2^k$, $b \in B$, $k \in \mathbb{N}$, and that $|W| = \max_{b \in B} r_b$ (similar arguments can be used for less regular rhythms and weeks). For a 2-week planning horizon, any solution $(B^1, B^2)$ for the MPSTDP-S has just 1 week-symmetric solution: $(B^2, B^1)$. For a 4-week planning horizon, a solution $(B^1, B^2, B^3, B^4)$ already has 7 week-symmetric solutions: $(B^3, B^2, B^1, B^4)$, $(B^1, B^4, B^3, B^2)$, $(B^3, B^4, B^1, B^2)$, $(B^2, B^1, B^4, B^3)$, $(B^4, B^1, B^2, B^3)$, $(B^2, B^3, B^4, B^1)$, and $(B^4, B^3, B^2, B^1)$. In general, each solution for the MPSTDP-S for a planning horizon of $|W| = 2^k$ weeks has $\prod_{l=1}^k 2^l - 1 = 2^{k(k+1)/2} - 1$ week-symmetric solutions. To see that, consider the case $|W| = 4$ and let $(B^1, B^2, B^3, B^4)$ be a solution. Moreover, let $(C^1, C^2, C^3, C^4)$, where $C^w = \{b \in B^w \mid r_b \leq 2\}$ is the set of customers with a weekly or biweekly rhythm. We obviously have $(C^1, C^2) = (C^3, C^4)$ as visiting weeks of the (bi-)weekly customers have to repeat every 2 weeks. As the 4-weekly customers are served in just one of the weeks, no permutation of the weeks will violate their week rhythm. Hence, to obtain all week-symmetric solutions for $(B^1, B^2, B^3, B^4)$, we have to consider: (*i*) all solutions symmetric to $(C^1, C^2)$, and (*ii*) all their cyclic permutations. Concerning (*i*), we only have $(C^2, C^1)$, giving us the two alternate solutions $(C^1, C^2, C^3, C^4)$ and $(C^2, C^1, C^4, C^3)$.

Concerning ($ii$), putting the 4-weekly customers back in, a cyclic permutation of $(C^1, C^2, C^3, C^4)$ results in the 4 week-symmetric solutions $(B^1, B^2, B^3, B^4)$, $(B^4, B^1, B^2, B^3)$, $(B^3, B^4, B^1, B^2)$, and $(B^2, B^3, B^4, B^1)$. A cyclic permutation of $(C^2, C^1, C^4, C^3)$ yields the other four. Thus, the total number of week-symmetric solutions for $|W| = 4$ is given by the number of week-symmetric solutions for $|W| = 2$ times the number of cyclic permutations, which equal the number of weeks in the planning horizon.

Unfortunately, both daily and weekly symmetries are problem inherent. In the formulation we presented above, this symmetry carries over to the formulation. This, however, does not necessarily have to be the case. We could, for example, think of a formulation where we do not have variables that assign individual customers to a day, but for each week we could define decision variables that select a set of five feasible day clusters (feasible in the sense that the day clusters do not violate the bounds on the daily service times and that each customer who is scheduled for this week is contained in one of the day clusters). This would, however, result in an exponential increase of the number of decision variables with respect to $|B|$.

We can think of different ways to try to reduce the number of symmetric solutions. The most simple and in some cases quite effective method is to pick a customer $b$ and a priori fix the first visit of $b$ to a specific week and day of the week, e.g., the first day of the first week. This can be done without loss of generality. By doing so, we reduce the number of day-symmetries with respect to the chosen week $w$ by a factor of $m$. Concerning weeks, if we make the same assumptions as above for the week rhythms and the length of the planning horizon, then fixing the first visit of customer $b$ with $r_b = 2^k$ to a given week $w \leq r_b$ reduces the number of week-symmetric solutions by a factor of $|W|$. To see why, consider again the discussion above. Fixing the first visiting week for $b$ rules out all symmetries due to cyclic permutations of solutions for the problem with $|W| = 2^{k-1}$ weeks, because for each week-symmetric solution of the latter problem, just one of the permutations will put $b$'s first visit in week $w$. The reduction in symmetric solutions is lower for $r_b < 2^k$ and, thus, we will always select a customer with $r_b = |W|$ in the remainder. We abbreviate this strategy as FRC.

Another possibility is to add constraints to formulation (MPSTDP-S) that sort the day clusters within each week by increasing customer indices. Let $B^w \subset B$ be a week cluster, $(B^{d^w}, \ldots, B^{d^w+m-1})$ its day clusters, where $d^w = \min_{d \in D^w} d$ is the first day in week $w$, and $b^d = \min_{b \in B^d} b$ is the customer with the smallest index in $B^d$, $d \in D^w$. Then we can rule out any day-symmetries with respect to $w$ by ensuring that $b^{d^w} < \ldots < b^{d^w+m-1}$. The corresponding constraints are given by

$$h_b^d \leq \sum_{b'=1}^{b-1} h_{b'}^{d-1} \qquad b = 2, \ldots, |B|, w \in W, d \in D^w \setminus \{d^w\}, \tag{7.2}$$

where we omit for simplicity the reference to the territories. We abbreviate this strategy as BDS.

## 7.3.4 Computational Experiments

Using the same data sets as in Sect. 7.3.1, the experiment results in Table 7.3 show the effect of fixing the first visit of a customer for the three problem variants. The structure of the table is almost identical to Table 7.2. For an easier comparison of the results, we also include a row "Deviation" that shows the deviation between the summarized results of Table 7.3 when compared to the respective summaries in Table 7.2. For the MIP gap and run times, we present the relative deviation, for the number of instances with proven optimal solution the absolute deviation. Looking at the two problem variants without days, we can see that the effect is negligible and may even result in higher gaps, worse run times, and fewer instances with a proven optimal solution, especially for the 30 customer data sets. For the scheduling subproblem MPSTDP-S the situation is however quite different. Across all week rhythm patterns, the average run times and/or the MIP gaps decrease and the total number of instances with a proven optimal solution increases.

Next, we evaluate the impact of adding the symmetry breaking constraints for days (7.2) to the formulation for the MPSTDP-S. The results for the 30 customer instances are shown in Table 7.4, whose structure is similar to the two previous tables. The reported deviations are with respect to the results in Table 7.2. Using strategy BDS leads to worse results when compared to Sect. 7.3.1, both in run times and MIP gaps. Combining both strategies, the results are inferior to just using strategy FRC. Hence, while strategy BDS helps to reduce symmetry "on paper," it is ineffective for our data sets. Unfortunately, other ideas that we tried, like sorting

**Table 7.3** Experiment result of using strategy FRC

| Pattern | MPSTDP w/o days | | | MPSTDP-S w/o days | | | MPSTDP-S | | |
|---|---|---|---|---|---|---|---|---|---|
| | Gap | Opt | Time | Gap | Opt | Time | Gap | Opt | Time |
| *30 customers* | | | | | | | | | |
| 1 | 0.01% | 5 | 15.4 | 0.00% | 5 | 0.5 | 0.01% | 5 | 146.4 |
| 2 | 1.81% | 3 | 778.1 | 0.00% | 5 | 1.0 | 2.29% | 3 | 1198.2 |
| 3 | 0.01% | 5 | 379.4 | 0.00% | 5 | 1.7 | 3.73% | 0 | 1800.5 |
| 4 | 1.75% | 4 | 661.4 | 0.00% | 5 | 5.9 | 14.70% | 0 | 1800.0 |
| 5 | 3.07% | 2 | 1500.4 | 0.00% | 5 | 5.4 | 9.00% | 0 | 1800.1 |
| 6 | 13.12% | 1 | 1620.4 | 0.00% | 5 | 17.1 | 12.39% | 0 | 1800.0 |
| **Summary** | **3.30%** | **20** | **825.9** | **0.00%** | **30** | **5.3** | **7.02%** | **8** | **1424.2** |
| Deviation | 86.8% | −5 | 26.0% | 0.00% | 0 | 3.9% | −17.0% | 2 | −4.9% |
| *40 customers* | | | | | | | | | |
| 1 | 0.01% | 5 | 42.5 | 0.00% | 5 | 0.6 | | | |
| 2 | 0.01% | 5 | 39.1 | 0.00% | 5 | 2.3 | | | |
| 3 | 0.01% | 5 | 482.4 | 0.00% | 5 | 3.8 | | | |
| 4 | 0.01% | 5 | 1026.4 | 0.00% | 5 | 12.5 | | | |
| 5 | 0.00% | 5 | 1799.9 | 0.01% | 5 | 10.9 | | | |
| 6 | 0.00% | 5 | 1417.1 | 0.00% | 5 | 72.8 | | | |
| **Summary** | **0.01%** | **30** | **757.5** | **0.00%** | **30** | **17.2** | | | |
| Deviation | −8.9% | 0 | 6.7% | 0.00% | 0 | −8.8% | | | |

**Table 7.4** Experiment result of using strategies BDS and FRC for the MPSTDP-S

| Rhythm | BDS | | | BDS and FRC | | |
|---|---|---|---|---|---|---|
| | Gap | Opt | Time | Gap | Opt | Time |
| 1 | 0.01% | 3 | 1052.7 | 0.01% | 3 | 208.5 |
| 2 | 9.98% | 0 | 1800.1 | 6.90% | 0 | 1800.1 |
| 3 | 11.22% | 0 | 1800.1 | 8.25% | 0 | 1800.1 |
| 4 | 27.46% | 0 | 1800.1 | 29.46% | 0 | 1800.3 |
| 5 | 12.88% | 0 | 1800.1 | 11.10% | 0 | 1800.1 |
| 6 | 17.32% | 0 | 1800.1 | 8.36% | 0 | 1800.1 |
| **Summary** | **12.4%** | **3** | **1725.4** | **10.3%** | **3** | **1535.2** |
| Deviation | 47.0% | −3 | 15.2% | 21.8% | −3 | 2.5% |

the service times of the weeks and of the days within each week in an increasing order, have proven to be equally ineffective.

## 7.3.5 Mathematical Formulations for Weekday Regularities

When scheduling visits, it might well be that the weekday when a customer is served changes from visit to visit. While some customers will not care, others might be unhappy with this. For those customers, we make sure that they are always served on the same day of the week. We call this requirement a *strict weekday regularity*. Enforcing a strict regularity will reduce the flexibility of planning. Even if customers prefer a strict regularity, they may not object if the service provider reschedules the visit to some other weekday every now and then. Thus, we also consider a hybrid version in which a regular weekday is fixed for serving a customer, but we are allowed to deviate from that given weekday for a predetermined number of visits. We call this a *partial weekday regularity*. For the ease of exposition, we focus on the scheduling subproblem MPSTDP-S. The corresponding modifications in formulation (MPSTDP) are straightforward. We start with the case of a strict weekday regularity.

### 7.3.5.1 Strict Weekday Regularities

Let $b \in B$ be a customer who requests to be visited always on the same weekday. Let $d_b^{first} \in \{1, \ldots, r_b m\}$ be the first visiting day of customer $b$. The set of all visiting days for $b$ is then given by $WD(d_b^{first}) = \{d \in D \mid d = d_b^{first} + k \cdot r_b m, k \in \mathbb{N}_0\}$. Note that $d_b^{first}$ is unique because of Constraints (7.1b) and (7.1j). We call $d_b^{first}$ the *regular weekday* of customer $b$. To model a strict regularity, we add the constraints

$$h_b^{d'} = h_b^d \qquad d' \in \{1, \ldots, r_b m\}, \ d \in DW(d'). \tag{7.3}$$

to formulation (MPSTDP-S).

A strict weekday regularity helps to reduce symmetry across weeks. Recalling the discussion in the previous section, consider a solution where the first visiting week, but not the first visiting day is the same for two customers $b$ and $b'$ with a strict regularity and identical week rhythms. While all permutations of days in the first week $w$ still yield day-symmetric solutions with respect to $w$, in all following visiting weeks $w + kr_b$, $k \in \mathbb{N}$, only permutations which schedule the visiting days of $b$ and $b'$ on the same days as in week $w$ are feasible. For the case of 5 weekdays, this reduces the number of day-symmetric solutions in subsequent visiting weeks from $5! = 120$ to $3! = 6$. And the smaller the week rhythms of the customers with strict regularities are, the larger the reduction will be. We refer to Sect. 7.4.1 for experiment results.

### 7.3.5.2 Partial Weekday Regularities

Let $b \in B$ be a customer who allows for a partial weekday regularity and $d_b \in \{1, \ldots, r_b m\}$ denote the regular weekday of $b$. By fixing the regular weekday in the first $r_b m$ weeks, the set of regular visiting days is again given by $WD(d_b)$. Let $f_b$ be the allowed number of visiting days different from that, i.e., for a solution of (MPSTDP-S), we must have $|\{d \in D \mid h_b^d = 1\} \setminus WD(d_b)| \le f_b$. To model these constraints, we define the decision variables

$$\hat{h}_b^{d_b} = \begin{cases} 1, & \text{if } d_b \text{ determines the regular weekday for customer } b \\ 0, & \text{otherwise,} \end{cases}$$

for $d_b \in \{1, \ldots, r_b m\}$. First, we have to select a regular weekday for customer $b$

$$\sum_{d_b=1}^{r_b m} \hat{h}_b^{d_b} = 1. \tag{7.4}$$

It is enough to determine the regular weekday within the first $r_b$ weeks. Observe that this only establishes the regular weekday for customer $b$; it does not imply that the first actual visit to $b$ has to be on this day. Second, we have to ensure that we deviate from the regular weekday at most $f_b$ times:

$$\sum_{d \in D \setminus WD(d_b)} h_b^d \le f_b \hat{h}_b^{d_b} + \frac{|W|}{r_b}(1 - \hat{h}_b^{d_b}) \qquad d_b \in \{1, \ldots, r_b m\} \tag{7.5}$$

For a given day $d$, the left-hand side counts the number of visits that do not match the regular weekday $d_b$. If $d_b$ is not selected as the regular weekday, then the right-hand side reduces to $|W|/r_b$, the total number of visits to $b$ during the planning horizon.

The partial weekday regularities still help to reduce symmetry across weeks, albeit to a smaller extent than for strict regularities. We refer to Sect. 7.4.1 for experiment results.

## 7.4 Solving the Scheduling Subproblem

As seen in the previous section, only very small data sets can be solved optimally with formulation (MPSTDP-S) within half an hour. However, typical problem instances often contain several hundred customers and span a time horizon of 24 or 48 weeks. In the following, we present a heuristic as well as a specially tailored branch-and-price algorithm for the MPSTDP-S.

### 7.4.1 A Heuristic Algorithm for the MPSTDP-S

In Bender et al. [1], we present a simple, yet efficient heuristic for solving the scheduling subproblem. This heuristic is based on Cooper's location-allocation method [8], which was first used in the context of (political) districting in the seminal paper of Hess et al. [11]. The idea of the heuristic is to decompose the simultaneous location and allocation decisions into two independent phases, a location phase and an allocation phase, which are then alternatingly performed until a satisfactory result is obtained.

Adapted to our problem, for a given set of week and day centers, the allocation phase determines a balanced and compact assignment of customers to week and day centers. This is done by fixing the values of the week and day center variables $x_i^w$ and $y_i^d$, respectively, in formulation (MPSTDP-S) and then solving it to obtain the optimal visiting weeks and days for each customer, as well as the corresponding week and day clusters. In the location phase, we take the set of week and day clusters derived from the allocation phase and determine an optimal center within each cluster. This can be done by simply finding the customer within each cluster $B' \subset B$ that minimizes the sum of distances to all customers of the cluster, i.e., finding $i^* = \arg\min_{i \in B} \sum_{b \in B'} c_{bi}$.

A crucial aspect of the heuristic is how to choose a "good" set of initial centers to kick off the heuristic. Hess et al. [11] pick initial district centers randomly. Recent approaches instead suggest the use of a k-means++ method to distribute the initial centers evenly across the planning area. While sounding very appealing, the latter approach does not work in our multi-period context, as the set of customers to be served changes from week to week (and day to day). In [1] we suggest to determine the week and day centers for the first $r^{min} \leq |W|$ weeks, and then copy those weeks recurringly to obtain initial centers for the remaining weeks and days of the planning horizon. $r^{min}$ is chosen as the smallest week rhythm of customers, i.e., $r^{min} = \min_{b \in B} r_b$. The motivation for this is based on the observation that if several customers with the same week rhythm $r'$ have the same first visiting week, then they will again be in the same week cluster $k \cdot r'$ weeks later, $k \in \mathbb{N}$. Hence, the corresponding week clusters will look—more or less—the same. Such similarities across weeks can only emerge after $r^{min}$ weeks, at the earliest. To find week centers for the first $r^{min}$ weeks, a modified k-means++ method is used to evenly distribute the week centers across the district. The modification affects the probability with which a customer is selected as center. A customer $b$ with week rhythm $r_b$ has to be served $|W|/r_b$ times in total. The smaller $r_b$, the more often $b$ is visited, and the greater the contribution of $b$ to the objective function is (in terms of distances). Hence, it would be beneficial to prefer those customers when choosing centers over customers with long week rhythms. Thus, we scale the probability of selecting a customer by $r_b$. To determine day centers, temporary week clusters for these first $r^{min}$ weeks are derived by assigning each customer, irrespective of her week rhythm and the weekly workload, to the closest week center. This will ensure that the day centers are close to their respective week centers. Then, the same procedure as for weeks is used to determine $m$ day centers for each temporary week cluster. With the week and day centers for the first $r^{min}$ weeks being given, the respective centers for a week $w > r^{min}$ are copied from week $w'$, where $w = w' + kr^{min}$, $k \in \mathbb{N}$. This finalizes the initialization step and we can enter into the location-allocation heuristic, starting with the allocation phase.

To assess the quality of the solutions provided by the location-allocation heuristic, we generated ten data sets each with 30 and 50 customers. The planning horizon for each is 4 weeks, with 5 days per week, and all week rhythms are powers of two. We use Gurobi 6.0.2 to solve formulation (MPSTDP-S) with a time limit of 10 h, using the location-allocation solution to warm-start the solver. All test are carried out on a machine with an Intel Xeon-2650 v2 CPU with 128 GB memory. Gurobi could solve none of the 50 customer instances to optimality and only three of the 30 customer instances. Comparing the heuristic solutions to the best incumbents from Gurobi, the former are on average 3% worse than the latter. Looking at the lower bounds provided by Gurobi upon termination, the average relative percentage deviation between those and the objective values of the location-allocation solutions is 8%. This underlines the high quality of the latter solutions, which can be computed on average in 4.6 s. For the three 30 customer data sets with known optimal solutions, the location-allocation solutions are 4.2%, 6.0%, and 7.3% worse than the optimal ones.

We also evaluate the practical implications of weekday regularities for 20 real-world data sets. The planning horizon for those experiments is 16 weeks, with 5 days per week. Each instance contains on average 115 customers, the visiting rhythms of all customers are powers of two, and the service times range from 22 to 42 min. We choose $\alpha_2 = \frac{1}{3}$ and $\alpha_3 = \frac{2}{3}$, putting a higher emphasis on daily compactness. Moreover, we set $LB^w = 0.85\mu^{week}$ and $UB^w = 1.15\mu^{week}$ as lower and upper bound, respectively, for the feasible range of weekly service times. Analogously, we define $LB^d$, $UB^d$, and $\mu^{day}$ for days, however with an allowed deviation of $\pm 0.3$ from the average. All test are carried out under Windows 7 on a machine with

an Intel Core i5-760 processor and 8 GB memory. The linear and integer programs
were solved using Gurobi 6.0.5.

To extend the analysis, we also considered week rhythms of $\{4, 8, 16\}$, still with
$|W| = 16$, and $\{3, 4, 6, 12, 16\}$ for a planning horizon of 48 weeks. In addition,
we considered a larger range of service times, going from 15 to 60, in steps of
5 min. We run the location-allocation heuristic for each data set and combination of
parameters, and recorded the run time of the heuristic. Moreover, to get an estimate
of the actual overall daily travel times of the service provider of a given solution,
we solved a symmetric traveling salesman problem for each day cluster's set of
customers, including the district center.

Starting with the run times, there is little difference between the original week
rhythms $\{1, 2, 4, 8, 16\}$ and the week rhythms $\{4, 8, 16\}$. The average run times
across all types of regularities (strict, partial, and none) for these two are 11 seconds
and 13 seconds, respectively. The average run time for the third set of week rhythms
however increases considerably to 77 s, which is likely due to the longer planning
horizon (48 vs. 16 weeks). Looking at the difference between the different types
of weekday regularities, partial regularities result in the longest run times, with an
average of 40 s across the three types of week rhythms, while instances with strict
regularities or no regularities at all only take around 22 s on average to solve.

Next, we evaluate the effect of regularities and the different types of week
rhythms on the actual daily travel times. When compared to the case of no
regularities, partial weekday regularities increase the average travel times by 0.59%
and 13.69% for the first and third set of week rhythms, respectively, and decrease
them by 0.17% for the second set of week rhythms. For strict regularities, the
increase for the three sets is, in order, 1.06%, 0.78%, and 18.51%. It becomes
apparent that weekday regularities for the third set of week rhythms, which is
much more irregular than the first two sets (which are all powers of two), have a
considerable effect on the daily travel times. In contrast to that for the first two
types, the impact of regularities on daily travel times is marginal.

Finally, we analyze if the range of service times has an effect on the daily
travel for different weekday regularities. Again, we use the daily travel times for
comparison and average over the three different patterns of week rhythms. For the
original service times, partial and strict regularities increase the travel times by
1.27% and 1.82%, respectively, over the case of no regularities. For the second type
of service times, these numbers increase to 6.08 and 8.88%. This indicates that an
increase in the range of service times is mirrored by a higher increase in the daily
travel times when weekday regularity is enforced.

## 7.4.2   A Branch-and-Price Algorithm for the MPSTDP-S

As only very small data sets can be solved optimally with standard MIP solvers, we
present in Bender et al. [2] a specially tailored branch-and-price algorithm. To start,
we first derive a set partitioning based reformulation of the scheduling subproblem.

Let $P(B)$ denote the power set of $B$. We define $S^{weeks} \subseteq P(B)$ as the set of all sets $s \in P(B)$ that are feasible week clusters (feasible in the sense that $s$ satisfies Constraints (7.1h) and (7.1i)). Analogously, we define $S^{days} \subseteq P(B)$ for days. For a cluster $s \in S^{weeks} \cup S^{days}$, we denote by $c_s = \min_{i \in B} \sum_{b \in s} c_{bi}$ the compactness of $s$. Moreover, we define the following decision variables:

$$\delta_s^w = \begin{cases} 1, & \text{if } s \in S^{weeks} \text{ is selected as week cluster in week } w \\ 0, & \text{otherwise.} \end{cases}$$

$$\gamma_s^d = \begin{cases} 1, & \text{if } s \in S^{days} \text{ is selected as day cluster on day } d \\ 0, & \text{otherwise.} \end{cases}$$

Then, an equivalent integer linear programming formulation for the MPSTDP-S is given as

(MPSTDP-S-SP)

$$\min \quad \alpha_2 \sum_{w \in W} \sum_{s \in S^{weeks}} c_s \delta_s^w + \alpha_3 \sum_{d \in D} \sum_{s \in S^{days}} c_s \gamma_s^d \tag{7.6a}$$

$$\text{s.t.} \quad \sum_{s \in S^{weeks}} \delta_s^w = 1 \qquad\qquad w \in W \tag{7.6b}$$

$$\sum_{w=1}^{r_b} \sum_{s \in S^{weeks}:b \in s} \delta_s^w = 1 \qquad\qquad b \in B \tag{7.6c}$$

$$\sum_{s \in S^{weeks}:b \in s} \delta_s^w = \sum_{s \in S^{weeks}:b \in s} \delta_s^{fv(b,w)} \qquad b \in B,\ w \in W,\ w > r_b \tag{7.6d}$$

$$\sum_{s \in S^{days}} \gamma_s^d = 1 \qquad\qquad d \in D \tag{7.6e}$$

$$\sum_{d \in D^w} \sum_{s \in S^{days}:b \in s} \gamma_s^d = \sum_{s \in S^{weeks}:b \in s} \delta_s^w \qquad b \in B,\ w \in W \tag{7.6f}$$

$$\delta_s^w \in \{0, 1\} \qquad\qquad s \in S^{weeks},\ w \in W \tag{7.6g}$$

$$\gamma_s^d \in \{0, 1\} \qquad\qquad s \in S^{days},\ w \in W \tag{7.6h}$$

The objective function (7.6a) represents a weighted average of the compactness of week and day clusters, where $\alpha_2, \alpha_3 \geq 0$ are weighting factors, see Sect. 7.3.1. Constraints (7.6b) select a cluster for each week. Constraints (7.6c) and (7.6d) ensure that each customer $b$ is served exactly once within the first $r_b$ weeks and every $r_b$ weeks after that, respectively. Constraints (7.6e) select a cluster for each day and Constraints (7.6f) make sure that a customer is included in one of the day clusters of each of her visiting weeks. The last two constraints specify the variable domains.

For our branch-and-price algorithm, we replace the set $S^{weeks}$ in formulation (MPSTDP-S-SP) by the week dependent sets $S^w \subseteq S^{weeks}$, $w \in W$. $S^w$ is the set of all week clusters which can be selected in week $w$. The reason for this is that in the course of the algorithm some of the generated clusters may only be valid for certain weeks, for example due to prior variable fixings. Analogously, we replace $S^{weeks}$ by the day dependent sets $S^d \subseteq S^{weeks}$, $d \in D$.

To start off the branch-and-price algorithm, we run the location-allocation heuristic and use the week and day clusters of that solution to create initial sets $S^w$ and $S^d$ of feasible week and day clusters, respectively. The resulting problem is called the restricted master problem (RMP). When solving the LP relaxation of the RMP, we check for negative reduced cost variables by solving the following pricing problem, where $\pi_0^w$, $\pi_1^w$, $\pi_2^{bw}$, $\pi_3^d$, and $\pi_4^{bw}$ denote the dual variables associated with Constraints (7.6b)–(7.6f):

$$(\text{S-PP}) \quad \min \quad \alpha_2 \sum_{b \in B} \sum_{i \in B} \sum_{w \in W} c_{bi} u_{bi}^w + \alpha_3 \sum_{b \in B} \sum_{i \in B} \sum_{d \in D} c_{bi} v_{bi}^d \tag{7.7a}$$

$$- \sum_{w \in W} \pi_0^w - \sum_{b \in B} \pi_1^b \sum_{i \in B} \sum_{w=1}^{r_b} u_{bi}^w$$

$$- \sum_{b \in B} \sum_{w=r_b+1}^{|W|} \pi_2^{bw} \sum_{i \in B} \left( u_{bi}^w - u_{bi}^{fv(w,b)} \right)$$

$$- \sum_{d \in D} \pi_3^d - \sum_{b \in B} \sum_{w \in W} \pi_4^{bw} \sum_{i \in B} \left( \sum_{d \in D^w} v_{bi}^d - u_{bi}^w \right)$$

$$\text{subject to} \quad u_{bi}^w \leq x_i^w \qquad\qquad i, b \in B, \ w \in W \tag{7.7b}$$

$$\sum_{i \in B} x_i^w = 1 \qquad\qquad w \in W \tag{7.7c}$$

$$\sum_{b \in B} \sum_{i \in B} s_b u_{bi}^w \leq UB^w \qquad\qquad w \in W \tag{7.7d}$$

$$\sum_{b \in B} \sum_{i \in B} s_b u_{bi}^w \geq LB^w \qquad\qquad w \in W \tag{7.7e}$$

$$v_{bi}^d \leq y_i^d \qquad\qquad i, b \in B, \ d \in D \tag{7.7f}$$

$$\sum_{i \in B} y_i^d = 1 \qquad\qquad d \in D \tag{7.7g}$$

$$\sum_{b \in B} \sum_{i \in B} s_b u_{bi}^d \leq U B^d \qquad d \in D \tag{7.7h}$$

$$\sum_{b \in B} \sum_{i \in B} s_b u_{bi}^d \geq L B^d \qquad d \in D \tag{7.7i}$$

$$x_i^w, \ y_i^d \in \{0, 1\} \qquad i \in B, \ d \in D, \ w \in W \tag{7.7j}$$

$$u_{bi}^w, \ v_{bi}^d \in \{0, 1\} \qquad b, i \in B, \ d \in D, \ w \in W \tag{7.7k}$$

Constraints (7.7b)–(7.7i) ensure that we generate feasible week and day clusters and, together with the first line of the objective function, that their compactness is computed correctly. The constraints are identical to the corresponding ones in formulation (MPSTDP), except for the missing district index $t$ and the variables $g_b^w$ and $h_b^d$, which have been replaced by $\sum_i u_{bi}^w$ and $\sum_i v_{bi}^d$, respectively (with $u_{bi}^w$ and $v_{bi}^d$ now being binary). As all constraints linking weeks across the planning horizon or linking visiting weeks with visiting days are included in the master and not the pricing problem, the latter decomposes into $|W| + |D|$ independent pricing problems, one for each week and each day. To solve each of the independent pricing problems, we break them further down by fixing a week center $i \in B$ and then solving the resulting formulation for the allocation variables $u_{bi}^w$; analogously for days. In that fashion we can identify up to $|B|(|W| + |D|)$ negatively priced variables, instead of just $|W| + |D|$ if we had solved the independent pricing problems directly. As the independent and broken down pricing problems are still NP-hard optimization problems, we first solve them with a Greedy heuristic. Only if the heuristic cannot find any negative reduced cost column, we solve the corresponding formulations exactly. At the end, all columns with negative reduced costs are added to the RMP.

Once the LP relaxation of the master problem has been solved to optimality in a given node of the branch-and-bound tree and we cannot fathom the node, we have to branch. As branching on variables of the restricted master problem changes the structure of the pricing problems, we decided to branch instead on the scheduling variables $g_b^w$ and $h_b^d$ of the original formulation (MPSTDP-S). For a fractional solution, we immediately get $g_b^w = \sum_{s \in S^w : b \in s} \delta_s^w$ and $h_b^d = \sum_{s \in S^d : b \in s} \gamma_s^d$. We prioritize branching on the week scheduling variables $g_b^w$. If all of them are integer, we branch on the day variables $h_b^d$. Whenever we fix any of those variables to 0 or 1, we remove all clusters from the RMP that would result in infeasible visiting weeks or days for the corresponding customers. If we have more than one fractional scheduling variable, we use pseudo cost branching to select a suitable one, see [2] for more details.

One of the main contributions in [2] is a symmetry-reduced branching (SRB) scheme that aims at eliminating as many week- and day-symmetric solutions as possible in the branching process. The idea of this scheme is to add additional variable fixings in a child node if we can guarantee that for any solution that becomes infeasible in this node due to the fixation, there exists a week- or day-symmetric solution in the other child node. Let us consider again our example from Sect. 7.3.3 with 4 weeks and week rhythms that are powers of 2. Assume that we have fixed the first visit of a customer $b' \in B$ with $r_{b'} = 4$ to the first day of the first week and that no other fixings exist so far. As a result, there are just two possible week-symmetric solutions remaining: $(B^1, B^2, B^3, B^4)$ and $(B^1, B^4, B^3, B^2)$. Consider now a customer $b \in B$ with $r_b = 4$ whose scheduling variable for week 2 is fractional, i.e., $0 < g_b^2 < 1$. Branching on $g_b^2$ generates two child nodes: $N_0$ with $g_b^2 = 0$ and $N_1$ with $g_b^2 = 1$. Considering node $N_0$, $b$ can still be served in week 1, 3, or 4. If we decide to serve $b$ in week 4, then we could as well have served her in week 2, due to week-symmetry. But then, this solution is already covered by node $N_1$. Hence, we can also fix $g_b^4 = 0$ in node $N_0$ without loss of generality. In an analogous fashion, we can add additional variable fixings for days.

### 7.4.3   Computational Experiments

In the following, we evaluate our algorithm using nine real-world data sets. The planning horizon consists of 4 weeks, with 5 days per week. The instances range from 25 to 35 customers and the service times from 10 to 330 min. Each customer has a power-of-two week rhythm $r_b \in \{1, 2, 4\}$. As before, we choose $\alpha_2 = 1/3$ and $\alpha_3 = 2/3$. Moreover, we set $LB^w = 0.9\mu^{week}$ and $UB^w = 1.1\mu^{week}$ for weeks and $LB^d = 0.8\mu^{day}$ and $UB^d = 1.2\mu^{day}$ for days. All tests are carried out under Ubuntu 16 on a machine with an Intel Xeon E5-2650 processor and 128 GB memory. The linear and integer programs are solved using Gurobi 7.0.1.

An excerpt of the results can be found in Table 7.5 (we refer to [2] for a more detailed analysis). The first column lists the number of the data set (for the purpose of referencing, we keep the numbering used in [2]). The next three columns present the results for solving formulation (MPSTDP-S) with a time limit of 10 h. The columns labelled $Gap$, $Obj$, and $Time$ show the MIP gap reported by Gurobi upon termination, the objective value of the best found solution, and the total runtime in seconds, respectively. The following two pairs of columns show the results for the branch-and-price algorithm, first without and then with our symmetry-reduced branching scheme. The objective value of optimal solutions is printed in bold.

We can see that even with a running time of 10 h, only four out of the nine instances can be solved to proven optimality using formulation (MPSTDP-S), although eight out of nine solutions are in fact optimal. While finding fewer optimal solutions, the branch-and-price algorithm can considerably reduce the runtimes, cutting the average by half. Finally, including the SRB in the branch-and-price algorithm, proven optimal solutions can be found for all nine instances, with the run times being two orders of magnitude smaller than the ones for (MPSTDP-S).

**Table 7.5** Results for the branch-and-price algorithm

| Instance | (MPSTDP-S) Gap | Obj | Time | B and P Obj | Time | B and P w/SRB Obj | Time |
|---|---|---|---|---|---|---|---|
| 1 | 0.01% | **1908.5** | 1132 | **1908.5** | 606 | **1908.5** | 54 |
| 2 | 1.91% | **1228.6** | 36,000 | **1228.6** | 8 | **1228.6** | 2 |
| 3 | 0.64% | **1893.7** | 36,000 | **1893.7** | 4284 | **1893.7** | 114 |
| 4 | 0.01% | **1702.5** | 12,493 | 1761.0 | 36,000 | **1702.5** | 458 |
| 5 | 0.17% | **2006.4** | 36,000 | **2006.4** | 33,995 | **2006.4** | 2133 |
| 8 | 0.01% | **2070.6** | 24,468 | **2070.6** | 171 | **2070.6** | 84 |
| 9 | 2.99% | 1949.1 | 36,000 | 1946.8 | 36,000 | **1946.6** | 416 |
| 10 | 1.27% | **1714.8** | 36,000 | **1714.8** | 1787 | **1714.8** | 29 |
| 11 | 0.00% | **2067.8** | 9844 | **2067.8** | 315 | **2067.8** | 16 |
| **Avg** | **0.78%** | | **25,326** | | **12,574** | | **365** |

## 7.5 Conclusions

In this chapter, we presented the multi-period service territory design problem. The problem extends classical districting problems by a scheduling component and is of high practical relevance for many industries, most notably in the retail and maintenance sector. We introduced a mixed-integer linear programming formulation for the MPSTDP and the preliminary experiments we carried out show that the problem is very challenging. One reason for this is the high amount of symmetry inherent in the MPSTDP, which carries over to our formulation. As the districting subproblem is very well studied, we focussed on the scheduling subproblem MPSTDP-S, proposing two new algorithms for it. The first is a location-allocation based heuristic which produces high-quality solutions within a short time. This heuristic outperformed the—then current—algorithm of PTV for the scheduling subproblem and has now replaced it. The second is a specially tailored branch-and-price algorithm which is able to decrease the run times by two orders of magnitudes when compared to the state-of-the-art solver Gurobi for small sized instances. This impressive improvement is mostly due to novel and very efficient symmetry-reduction techniques in the branching process.

## References

1. Bender, M., Meyer, A., Kalcsics, J., Nickel, S.: The multi-period service territory design problem–an introduction, a model and a heuristic approach. Transport Res. E-Log Transport Rev. **96**, 135–57 (2016)
2. Bender, M., Kalcsics, J., Nickel, S., Pouls, M.: A branch-and-price algorithm for the scheduling of customer visits in the context of multi-period service territory design. Eur. J. Oper. Res. **269**(1), 382–396 (2018)

3. Benzarti, E., Sahin, E., Dallery, Y.: Operations management applied to home care services: analysis of the districting problem. Decis. Support. Syst. **55**(2), 587–598 (2013)
4. Blais, M., Lapierre, S., Laporte, G.: Solving a home-care districting problem in an urban setting. J. Oper. Res. Soc. **54**(11), 1141–1147 (2003)
5. Blakeley, F., Argüello, B., Cao, B., Hall, W., Knolmajer, J.: Optimizing periodic maintenance operations for Schindler Elevator corporation. Interfaces **33**(1), 67–79 (2003)
6. Bodin, L., Levy, L.: The arc partitioning problem. Eur. J. Oper. Res. **53**(3), 393–401 (1991)
7. Butsch, A., Kalcsics, J., Laporte, G.: Districting for arc routing. INFORMS J. Comput. **26**(4), 809–824 (2014)
8. Cooper, L.: Location-allocation problems. Oper. Res. **11**, 331–343 (1963)
9. Hanafi, S., Fréville, A., Vaca, P.: Municipal solid waste collection: an effective data structure for solving the sectorization problem with local search methods. Inf. Syst. Oper. Res. **37**(3), 236–254, (1999)
10. Handley, L., Grofmann, B. (eds.): Redistricting in Comparative Perspective. Oxford University Press, New York (2008)
11. Hess, S.W., Weaver, J.B., Siegfeldt, H.J., Whelan, J.N., Zitlau, P.A.: Nonpartisan political redistricting by computer. Oper. Res. **13**(6), 998–1008 (1965)
12. Kalcsics, J.: Districting problems. In: Laporte, G., Nickel, S., Saldanha da Gamma, F. (eds.) Location Science, chap. 23, pp. 595–622. Springer, Cham (2015)
13. Kalcsics, J., Nickel, S., Schröder, M.: Towards a unified territorial design approach— applications, algorithms and GIS integration. Top **13**(1), 1–56 (2005)
14. Lei, H., Laporte, G., Liu, Y., Zhang T.: Dynamic design of sales territories. Comput. Oper. Res. **56**, 84–92 (2015)
15. Lei, H., Wang, R., Laporte, G.: Solving a multi-objective dynamic stochastic districting and routing problem with a co-evolutionary algorithm. Comput. Oper. Res. **67**, 12–24 (2016)
16. Lin, H.Y., Kao, J.J.: Subregion districting analysis for municipal solid waste collection privatization. J. Air Waste Manage. Assoc. **58**, 104–111 (2008)
17. Mourgaya, M., Vanderbeck, F.: Column generation based heuristic for tactical planning in multi-period vehicle routing. Eur. J. Oper. Res. **183**(3), 1028–1041 (2007)
18. Muyldermans, L., Cattrysse, D., Van Oudheusden, D., Lotan T.: Districting for salt spreading operations. Eur. J. Oper. Res. **139**(3), 521–532 (2002)
19. Ríos-Mercado, R.Z., López,-Pérez J.F.: Commercial territory design planning with realignment and disjoint assignment requirements. Omega **41**, 525–535 (2013)
20. Salazar-Aguilar, M.A., Ríos-Mercado, R.Z., Cabrera-Ríos, M.: New models for commercial territory design. Netw. Spat. Econ. **11**(3), 487–507 (2011)
21. Zoltners, A.A., Sinha, P.: Sales territory design: thirty years of modeling and implementation. Mark. Sci. **24**(3), 313–331 (2005)

# Chapter 8
# Designing Ambulance Service Districts Under Uncertainty

Shakiba Enayati, Osman Y. Özaltın, and Maria E. Mayorga

**Abstract** For ambulances, quick response to a medical emergency is critical. Limiting response area for each ambulance may lead to shorter response times to emergency scenes and more evenly distributed workload for ambulances. We propose a two-stage stochastic mixed-integer programming model to address the service district design problem under uncertainty. The proposed model recommends how to locate ambulances to the waiting sites in the service area, and how to assign a set of demand zones to each ambulance at different backup levels. Our proposed Stochastic Service District Design (SSDD) model enables quick response times by jointly addressing the location and dispatching policies in a stochastic and dynamic environment. Each backup level is associated with a given response time threshold. The objective function is to maximize the expected number of covered calls while restricting the workload of each ambulance. The proposed model can be optimized offline as is commonly done for "patrol beats" used in policing models. We evaluate the implementation of the proposed model via a discrete-event simulation, and compare the model with two baseline policies. Our computational results show a significant improvement in mean response time, reduction of 2 min, and statistically lower average workload of ambulances, of 4% on average, when the proposed model is fully implemented.

S. Enayati
State University of New York, Plattsburgh, NY, USA
e-mail: senay002@plattsburgh.edu

O. Y. Özaltın · M. E. Mayorga (✉)
North Carolina State University, Raleigh, NC, USA
e-mail: oyozalti@ncsu.edu; memayorg@ncsu.edu

© Springer Nature Switzerland AG 2020
R. Z. Ríos-Mercado (ed.), *Optimal Districting and Territory Design*, International
Series in Operations Research & Management Science 284,
https://doi.org/10.1007/978-3-030-34312-5_8

153

## 8.1   Introduction

Emergency medical service (EMS) refers to the provision of critical medical care outside of hospitals and transferring of patients to hospitals by equipped vehicles under medical supervision [27]. Each EMS agency operates in a specific service area to respond to emergency calls from the population within its geographic boundaries (i.e., town, city, or county). To better manage demand, the service area may be divided into smaller demand zones [15, 53]. Depending on the regulations and policies of the agency, demand zones may include a number of potential locations, i.e., waiting sites, for the vehicles, medical equipment, and paramedics to be stationed while at rest.

Emergency medical calls are routed through a dispatch center. Dispatchers answering calls interface with a computer-aided dispatch (CAD) system to collect information (e.g., call type, time of dispatch, time of arrival, time the vehicle left the scene and arrived at the hospital, and time the vehicle left the hospital and reported back to service). The service time captures the duration between vehicle dispatch to a call and vehicle return to service. The response time is the difference between time the unit is dispatched and its arrival to the scene.

Quick response to a medical emergency by an ambulance is critical. For example, the survival probability of cardiac arrest patients increases by 50% when attended to by an ambulance within 8 min versus 15 min [51], and it is estimated that survival rate decreases by 5.5% per min that the patient is without treatment [31]. Thus, reducing emergency response times is one of the main ways that EMS systems can decrease mortality of out-of-hospital patients.

In many communities, demand and traffic congestion continue to grow but resources cannot keep pace [25]. Under tight resource constraints, improving operational efficiency is the only way that most EMS systems will be able to maintain or improve quality of service and potentially save more lives. This is a rather difficult task; however, due to several complexities of EMS systems, they are geographically distributed, operate under uncertainty (e.g., calls arrive randomly and service times are stochastic) with heterogeneous customers (e.g., patients with different medical needs), and have inherent interdependencies (e.g., dispatching an ambulance for service impacts the service of future calls).

Due to such uncertain and complex system dynamics, EMS managers face many challenges at the strategic and tactical levels. At the strategic level, they need to decide what locations are potentially appropriate for ambulance waiting sites. This problem has been studied in the literature as a facility location problem [13, 14, 19, 40, 48]. Moreover, it has been shown that limiting or adapting the response area may decrease the average EMS system response time [43] and improve personnel efficiency and satisfaction. Thus, at the strategic level, EMS management may need to investigate the advantages of dividing the original service area into several predefined response districts so as to improve performance of the EMS system. This problem has been handled in the literature as a districting problem [21, 32, 43].

At the tactical level, there are challenges regarding dispatching policies and redeployment of ambulances. A dispatching policy is a set of rules for selecting proper ambulances to respond to emergency calls. This problem has been extensively covered in the literature [33, 44–46]. Ambulance redeployment refers to a strategy that potentially helps to improve EMS performance by repositioning idle ambulances to compensate for unavailability of busy ambulances. Also, redeployment may refer to finding the next location of busy ambulances after service completion (instead of sending ambulances back to their home stations). Many approaches have been proposed in the literature for the redeployment of ambulances [41, 42, 58–60]. Some studies also handle integrated problems in one approach, such as joint location and dispatching problems [23, 66] or joint redeployment and dispatching problems [18].

This chapter concentrates on the development and the evaluation of a two-stage stochastic programming model to address the service district design problem. The proposed Stochastic Service District Design (SSDD) model enables quick response times by jointly addressing the location and dispatching policies in a stochastic and dynamic environment. That is, it first locates ambulances at the potential waiting sites in the service area. Each ambulance is then assigned to a set of demand zones to respond to their emergency calls at different backup levels. Each backup level corresponds to a response time threshold. The proposed SSDD model can be optimized *offline* and the outputs regarding the dispatching policy are easily applied in a dynamic manner. Such an implementation is similar to the idea of "patrol-beats" used in policing models and it reduces the technological capabilities required to integrate the model within basic computer-aided dispatch (CAD) systems.

In the remainder of this chapter, we briefly review the related literature in Sect. 8.2. We present the model formulation in Sect. 8.3. We discuss the output representation and the numerical results of implementing the proposed model via simulation in Sect. 8.4. We finally conclude in Sect. 8.5.

## 8.2 Literature Review

EMS providers seek to minimize response times and maximize coverage, i.e., the proportion of calls answered within a given time threshold. The EMS literature has addressed four different problems with the goal of improving service: ambulance location/allocation, redeployment, and dispatching problems as well as the districting problem. These problems are formulated and solved in three (not necessarily mutually exclusive) streams: static/deterministic, dynamic/real-time, and probabilistic. Earlier approaches mostly dealt with static and deterministic models for location/allocation problems [13, 65]. Static models are proposed for the strategic decisions in the planning stage [10]. Dynamic and real-time approaches address tactical decision such as ambulance redeployment [27]. Probabilistic approaches handle the uncertainty of ambulance availability. Brotcorne et al. [10], Goldberg [22], Li et al. [37], Başar et al. [5], Bélanger et al. [7], and Aringhieri et al. [4] reviewed mathematical models in the EMS literature. The remainder of this section reviews studies addressing ambulance location/allocation as well as ambulance dispatching and redeployment, with a focus on those works that address uncertainty. We also summarize the studies on service districting problem in the EMS literature as well as in the literature of other applications.

## 8.2.1 Ambulance Location/Allocation

While there is a large body of research related to resource location/allocation in EMS systems, there is a lack of mathematical models that explicitly consider uncertainty. The maximum expected covering location problem (MEXCLP) was formulated as one of the first probabilistic models for the ambulance location problem [14]. MEXCLP assumed that each ambulance has the same probability, called the *busy fraction*, of being unavailable to respond to a call, and all ambulances are independent. Employing the busy fraction concept, the maximum availability location problem was proposed to maximize the demand covered with a given probability [54]. Several studies viewed ambulances as servers in a queuing system [6, 8, 11, 39, 40], and applied the hypercube model of Larson [32] to estimate busy fractions associated with the whole system or with a specific waiting site. Furthermore, compliance tables are designed to prescribe ambulance allocation given a certain fleet size. A Markov model was proposed to estimate the busy probabilities of ambulances for a given compliance table [1]. Optimizing a compliance table design was first proposed by Sudtachat et al. [63]. Another way to deal with uncertainty is through detailed simulation models, such as Andersson and Varbrand [2] and Sudtachat et al. [62], but these can only be used to compare proposed policies or heuristics, not to find optimal solutions, and cannot be generalized to systems under different operating conditions.

## 8.2.2 Ambulance Dispatching and Redeployment

Advances in CAD systems and access to data have spurred research in real-time models that are executed *online*. These models focus on dispatching and redeploying ambulances based on the current system status and demand forecasts [41, 53, 58, 59]. A mixed-integer program was formulated to provide a dynamic relocation strategy that maximizes the expected demand coverage while controlling the number of waiting site relocations [20]. System-status management models consider real-time decisions and not strategic decisions. Real-time recommendations must still adhere to the underlying design and response policies [42]. Advanced CAD systems may be linked with GIS technology to make dispatch recommendations based on real-time traffic information and demand forecasts. However, suburban and rural EMS systems typically have basic CAD systems that lack GIS capabilities. Even with advanced CAD systems, enhancing their capabilities is expensive and some agencies view this as an avoidable cost [56]. Thus, it is important to provide guidelines for data-driven optimization models that can be integrated into existing CAD systems.

## 8.2.3   Service District Design

Service district design is a strategic decision that is made *offline* for constructing the primary (and possibly backup) response area(s) of ambulances to improve coverage and enable quick response times. This problem can be addressed from two complementary perspectives. The first perspective involves grouping demand zones into compact and contiguous districts such that the set of assigned vehicles to each district is primarily responsible for responding to calls within that district. This is the traditional districting design approach which creates clear physical borders for districts. In geographically compact districts the distance to the center does not vary widely, and contiguity ensures that it is possible to travel between any two points within a district without having to leave the district. A mixed-integer linear program was formulated to optimize weighted objectives of district balance (defined as relative deviation of the district size from the average district size), and recommend compact and contiguous districts for political districting, police patrol area delineation, and sales territory design applications [30]. See additional examples in the literature of political and public service districting problems [12, 47, 64]. Modeling such a districting design would result in highly symmetric mathematical programs in which several equivalent solutions can be obtained by changing the "center" of a district plan to any of the demand zones included in the district. Therefore, solution algorithms developed for this modeling perspective have to explore many alternative symmetric solutions which increases the computational burden [61]. The problem of dividing an area into a set of compact and contiguous districts is NP-complete [52]. The existing models and solution approaches in the literature are limited to solve only small instances [16, 57], or some common districting criteria (i.e., contiguity, compactness, balanced districts) are not fully enforced in the formulation [36, 50, 55]. Considering uncertainty in the EMS systems is even more challenging in such districting models. The second perspective to district design assigns a set of demand zones to each vehicle at different backup levels. This approach differs from designing contiguous and compact districts as it prescribes how to dispatch vehicles to calls in real time without explicitly partitioning the response area into geographical districts. This chapter adopts the latter perspective to address the districting problem in EMS systems with uncertainty considerations.

There have been few works in the literature related to district design of EMS systems. An optimization framework was developed using genetic algorithms to district a region and deploy ambulances on highways to minimize the average response time; however, the results were limited to a linear service region [26]. A heuristic was proposed to design districts for an EMS system, and developed a simulation model to evaluate the performance of several integrated districting and dispatching policies in terms of patient survival probability [43]. A mixed-integer program was formulated to locate and dispatch ambulances through district design [3]. They consider the uncertainty in ambulance availability and travel times through a hypercube model approximation.

The districting problem has been extensively studied in various other applications such as geography, politics, ecology, business, and public service. A broad perspective on common criteria and limitations in districting applications was provided [28]. Not many works in the literature have studied stochastic districting problems. A two-stage stochastic program was developed for vehicle routing and districting problem with stochastic customers in which districting decisions are made in the first stage and the routing expenses are approximated in the second stage [34]. They addressed the compactness of districts in the objective function and developed a large neighborhood search heuristic to solve the model. Also, the districting and routing problem were addressed with stochastic demand using a two-stage stochastic program and developed Tabu search and multistart heuristics [24].

The proposed stochastic programming model in this chapter is the first analytic model that explicitly considers uncertainty for the EMS service district design problem. The proposed formulation can model complex EMS system dynamics while leveraging historical data. Only a few papers considered stochastic programming formulations for EMS problems. A two-stage stochastic programming model was developed that is solved repeatedly on a rolling horizon basis in real time to minimize the number of relocation moves during the planning horizon, while maintaining an acceptable service level [49]. Furthermore, a two-stage stochastic programming model with chance constraints was formulated to locate ambulances in the first stage, and allocate demand points to ambulances in the second stage [8]. However, their proposed model does not consider the uncertainty in the availability of ambulances when evaluating coverage. In a similar vein, a chance-constrained optimization to relocate ambulances was proposed [38]. In summary, there is a lack of mathematical models for spatially distributed service systems that explicitly consider uncertainty which is an endemic feature of such systems.

## 8.3    Model Formulation

Better service district designs may lead to shorter response times, more equitable division of the workload, familiarization with the assigned area, and more efficient use of personnel. The average response time and the variation of workload among different districts are two important performance measures of service district design. Limiting the primary response area of ambulances enables the EMS system to decrease the average response time of paramedic support to the scene [43]. However, a service district design, just like any other decision for improving operational efficiency in EMS systems, has to consider uncertainty as EMS systems deal with uncertain demand with several stochastic attributes such as arrival time, location, and service time of emergency calls [18].

We generate a finite set of scenarios to include the uncertainty of the emergency call sequence into the decision-making process. Each scenario consists of a set of calls that occurred during a shift and call specifications, i.e., location, arrival time, and service time. We assume that the likelihood of an ambulance being unavailable to respond to a call (i.e., busy fraction) is identical and independent for all ambulances under each scenario [14].

We formulate a two-stage stochastic programming model to design EMS service districts with the goal of maximizing expected number of emergency calls that are responded to in a timely manner while limiting the workload of ambulances. Each district consists of multiple demand zones, and the ambulance assigned to a district is responsible to serve calls in those demand zones within a certain response time threshold. In the first stage, ambulances are assigned to waiting sites, and demand zones are assigned to ambulances at different backup levels. Each backup level is associated with a response time threshold. The first-level backup ambulance of a demand zone primarily responds to calls from that zone; however, if the first-level backup ambulance is busy, then the second-level backup ambulance responds, and so on. The second-stage problem measures the expected proportion of calls that are served by each ambulance at all backup levels as a result of the first-stage assignments. The proposed model in fact prescribes how to dispatch ambulances to emergency calls depending on their real-time availability at different backup levels without explicitly partitioning the service area into compact and contiguous districts.

Let $\mathcal{I}$ be the set of demand zones, $\mathcal{J}$ be the set of waiting sites ($\mathcal{J} \subseteq \mathcal{I}$), and $\mathcal{R}$ be the set of backup levels, i.e., $\{1, \ldots, R\}$. We define $\mathcal{J}_{ir}$ as the set of waiting sites that can serve demand zone $i \in \mathcal{I}$ at the $r^{th} \in \mathcal{R}$ backup level (i.e., $\mathcal{J}_{ir}$ includes those waiting sites that are located within the given time threshold defined for the $r^{th}$ backup level from demand zone $i$). The number of available ambulances is denoted by $\kappa$. The first-stage decision variables are denoted by:

- $x_j \in \{0, 1\}$: equals 1 if an ambulance is located at site $j \in \mathcal{J}$, and 0 otherwise.
- $y_{ijr} \in \{0, 1\}$: equals 1 if the $r^{th} \in \mathcal{R}$ backup ambulance for demand zone $i \in \mathcal{I}$ is located at site $j \in \mathcal{J}$, and 0 otherwise.

Let $\mathcal{X} := \{x_j, j \in \mathcal{J}\}$ and $\mathcal{Y} := \{y_{ijr}, i \in \mathcal{I}, j \in \mathcal{J}, r \in \mathcal{R}\}$. The Stochastic Service District Design (SSDD) model is then formulated as:

$$\text{(SSDD)} \quad Z^* = \max \mathrm{E}\left(Q_s(\mathcal{X}, \mathcal{Y})\right) \tag{8.1a}$$

$$\text{subject to} \quad \sum_{j \in \mathcal{J}_{ir}} y_{ijr} \geq 1 \qquad i \in \mathcal{I}, r \in \mathcal{R} \tag{8.1b}$$

$$\sum_{r \in \mathcal{R}} y_{ijr} \leq x_j \qquad i \in \mathcal{I}, j \in \mathcal{J} \tag{8.1c}$$

$$\sum_{j \in \mathcal{J}} x_j \leq \kappa \tag{8.1d}$$

$$x_j, y_{ijr} \in \{0, 1\} \qquad i \in \mathcal{I}, j \in \mathcal{J}, r \in \mathcal{R} \tag{8.1e}$$

The first-stage objective function (8.1a) maximizes the expected recourse function which will be elaborated in the second-stage problem. Constraints (8.1b) state that demand zone $i \in \mathcal{I}$ must be covered by at least one ambulance located at covering site $j \in \mathcal{J}_{ir}$ at backup level $r \in \mathcal{R}$. Constraints (8.1c) ensure that site $j \in \mathcal{J}$ responds to calls from demand zone $i \in \mathcal{I}$ at any backup level $r \in \mathcal{R}$ only if there is an ambulance located at site $j \in \mathcal{J}$. Constraint (8.1d) enforces the fleet size limit.

Let $\mathcal{S}$ be the set of generated scenarios. The second-stage decision variable is defined as follows:

- $z_{ij}^s \in [0, 1]$: the proportion of calls from demand zone $i \in \mathcal{I}$ that are served by an ambulance at waiting site $j \in \mathcal{J}$ under scenario $s \in \mathcal{S}$.

Table 8.1 summarizes the notation of parameters used in the second-stage problem of the SSDD model. The second-stage problem under scenario $s \in \mathcal{S}$ is given by:

$$Q_s(\mathcal{X}, \mathcal{Y}) = \max \sum_{i \in \mathcal{I}} \sum_{j \in \mathcal{J}} d_i^s z_{ij}^s \tag{8.2a}$$

$$\text{subject to} \quad z_{ij}^s \leq \sum_{r \in \mathcal{R}} \rho_{ir}^s y_{ijr} \qquad i \in \mathcal{I}, j \in \mathcal{J} \tag{8.2b}$$

$$\sum_{i \in \mathcal{I}} d_i^s z_{ij}^s \leq F^{\max} x_j \quad j \in \mathcal{J} \tag{8.2c}$$

$$z_{ij}^s \geq 0 \qquad i \in \mathcal{I}, j \in \mathcal{J} \tag{8.2d}$$

The second-stage objective function (8.2a) maximizes the coverage. Constraints (8.2b) compute the proportion of calls in demand zone $i \in \mathcal{I}$ that are

**Table 8.1** Parameters in the second stage of the service district design model

| | |
|---|---|
| $\rho_{ir}^s$ | Expected proportion of calls from demand zone $i$ under scenario $s$ that arrive when the first $(r-1)$ backup ambulances are busy and the $r^{th}$ backup ambulance is available: $\sum_{r \in \mathcal{R}} \rho_{ir}^s \leq 1, i \in \mathcal{I}, r \in \mathcal{R}$, and $s \in \mathcal{S}$ |
| $d_i^s$ | Ratio of calls from demand zone $i \in \mathcal{I}$ to all calls during the planning horizon under scenario $s \in \mathcal{S}$ |
| $F^{\max}$ | Maximum proportion of all calls that can be responded by each ambulance |

responded to from site $j \in \mathcal{J}$ at different backup levels. Note that parameter $\rho_{ir}^{s}$ is a function of the busy fraction probability of ambulances and the sequence of calls generated from a demand zone under each scenario. Constraints (8.2c) establish the upper bound on the proportion of all calls that can be responded to by site $j \in \mathcal{J}$ at which an ambulance is located. We estimate parameter $F^{\max}$ in constraints (8.2c) by:

$$F^{\max} = \frac{\max_{s \in \mathcal{S}} (\sum_{i \in I} d_i^s)}{\kappa} \tag{8.3}$$

Equation (8.3) sets the upper bound for the workload of each ambulance by equally distributing the calls between ambulances, i.e., we divide the maximum total number of calls during a shift generated in scenario set $\mathcal{S}$ by the number of available ambulances. Finally, constraints (8.2d) enforce the second-stage decision variables to be nonnegative.

## 8.4   Computational Results

We discuss the model output on two different instance sets:

1. Small instances: configuration of a service area is randomly generated. Emergency calls are uniformly distributed in the service area and their arrival times and service times are also generated randomly from a uniform distribution. All demand zones are considered as potential waiting sites for ambulances ($|I| = |J|$).

2. Large instance: scenarios are generated based on a real dataset collected from an EMS agency in Mecklenburg County, North Carolina. The service area is about 540 square miles that is hypothetically divided into 168 demand zones (i.e., $|I| = 168$). Each zone is 4 square miles ($2 \times 2$) [17]. The set of waiting sites $J$ is assumed to be the centroid of each demand zone (i.e., $|J| = 168$) because ambulances are allowed to wait at any location in Mecklenburg County. The dataset includes about 50,000 calls in Mecklenburg County for 650 shifts in 2004. The spatial demand distribution is highly concentrated in the central demand zones and is more sparse in the peripheral areas. Furthermore, demand significantly varies based on day of the week and time of the day [17]. We assume that each shift in the data corresponds to a scenario.

In both instances, we consider three different backup levels, $|R| = 3$, with corresponding coverage thresholds of 8, 12, and 16 min. We solve the extensive form of the proposed SSDD model (8.1). All computational experiments have been performed using CPLEX 12.6 on a Windows computer with 32 GB of memory and 3.4 GHz Intel Core i7 processor. Section 8.4.1 presents the model outputs on small examples. Section 8.4.2 evaluates the solution of the large instance via simulation.

### 8.4.1 Discussion of the Model Output

The SSDD model outputs the optimal waiting sites for available ambulances. Also, it determines the response design at all backup levels corresponding to each ambulance site. We present the model outputs on two different small examples in Figs. 8.1 and 8.2. In each figure, demand zones with asterisks indicate the waiting site at which an ambulance is located. Demand locations shown in dark gray, light gray, and white illustrate the zones for which the ambulance located at the asterisked site serve as the first, second, and third backup, respectively. Demand zones shown in red are not covered within any backup level of the corresponding ambulance at the asterisked site. The service area in both examples is a random subset of zones selected from the service area of the EMS agency at Mecklenburg County.

The first small example is shown in Fig. 8.1 which is the ambulance response design for a service area including 15 demand zones (i.e., $|I| = 15$), 4 ambulances (i.e., $\kappa = 4$), and 3 backup levels (i.e., $|R| = 3$). Figure 8.1a–d shows the response design at each backup level for ambulances stationed at waiting sites 3, 7, 9, and 11, respectively. This result is the optimal solution of the SSDD model for 200 scenarios in which the busy fraction of ambulances ranges between 41.6% and 64.3% during a 6-h shift. In this instance, the computational time to solve the extensive form of the SSDD model by CPLEX is 20 s. Figure 8.1 demonstrates that all demand zones are covered at least once within each backup level.

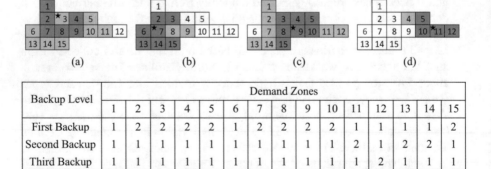

|                | Demand Zones |   |   |   |   |   |   |   |   |    |    |    |    |    |    |
|----------------|:---:|:---:|:---:|:---:|:---:|:---:|:---:|:---:|:---:|:---:|:---:|:---:|:---:|:---:|:---:|
| Backup Level   | 1 | 2 | 3 | 4 | 5 | 6 | 7 | 8 | 9 | 10 | 11 | 12 | 13 | 14 | 15 |
| First Backup   | 1 | 2 | 2 | 2 | 2 | 1 | 2 | 2 | 2 | 2 | 1 | 1 | 1 | 1 | 2 |
| Second Backup  | 1 | 1 | 1 | 1 | 1 | 1 | 1 | 1 | 1 | 1 | 2 | 1 | 2 | 2 | 1 |
| Third Backup   | 1 | 1 | 1 | 1 | 1 | 1 | 1 | 1 | 1 | 1 | 1 | 2 | 1 | 1 | 1 |
| Total Weighted | 6 | 9 | 9 | 9 | 9 | 6 | 9 | 9 | 9 | 9 | 8 | 7 | 8 | 8 | 9 |

**Fig. 8.1** Ambulance response design for a small example with 15 demand zones, 4 ambulances, and 3 backup levels. Demand zones with asterisks indicate the station at which an ambulance is located. Dark gray and light gray demand zones are covered within the first and the second backup levels by the corresponding ambulance located at the asterisked site, respectively. White zones demonstrate the third backup level. Demand zones shown in red are not covered within any backup level of the corresponding ambulance at the asterisked site. The table shows the number of ambulances covering at each backup level as well as the total weighted coverage for each demand zone. The first, second, and third backup levels weigh 3, 2, and 1, respectively

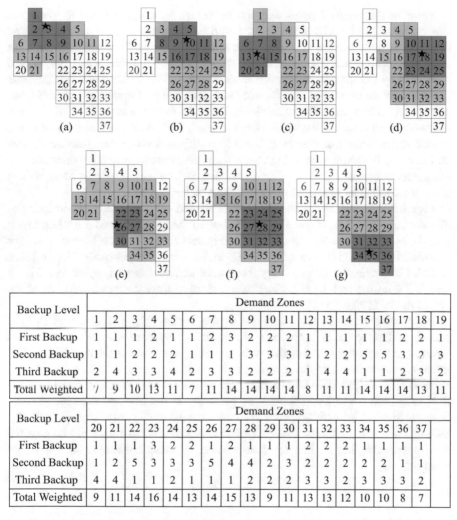

**Fig. 8.2** Ambulance response design for small example with 37 demand zones, 7 ambulances, and 3 backup levels. Demand zones with asterisk indicate the station at which an ambulance is located. Dark gray and light gray demand zones are covered within the first and the second backup levels by the corresponding ambulance located at the asterisked site, respectively. White zones demonstrate the third backup level. Demand zones shown in red are not covered within any backup level of the corresponding ambulance at the asterisked site. The table shows the number of ambulances covering at each backup level as well as the total weighted coverage for each demand zone. The first, second, and third backup levels weigh 3, 2, and 1, respectively

Some of the demand zones may not be served by an ambulance at a specific site as shown in Fig. 8.1d in red fonts, i.e., demand zones 1 and 6 are not covered by the ambulance located at station 11 at any backup level. However, this does not mean that those demand zones are not covered at all. According to Fig. 8.1, zone 6 is covered by the ambulances located at sites 3, 7, and 9. Furthermore, it is possible that the ambulance located at a site does not cover the corresponding demand zone at the first backup level. For example, in Fig. 8.1a the ambulance located at site 3 covers demand zone 3 at the second backup level. This can happen as a result of the workload constraint and may be difficult to justify in practice. Such circumstances do not occur frequently according to our trial runs, thus we employ a correction to the model output and assume if an ambulance is located at a site, it must always cover the corresponding demand zone at the first backup level.

Figure 8.2 illustrates the solution to the second small instance which includes 37 demand zones (i.e., $|I| = 37$), 7 ambulances (i.e., $\kappa = 7$), and 3 backup levels (i.e., $|R| = 3$). Figure 8.2a–g shows the response design at each backup level for ambulances stationed at sites 3, 10, 14, 18, 26, 28, and 35, respectively. This instance includes 200 scenarios and the busy fraction of ambulances ranges between 51.2% and 75.3% during an 8-h shift. The time required to solve the extensive form of the SSDD model is 72 s.

As demonstrated via two examples in this section, the SSDD model (8.1) is flexible to find the optimal solution to different EMS systems with different demand congestion levels and shift lengths. A drawback of our model is the possibility of an infeasible solution due to the constraints (8.1b) that force the system to cover each demand zone at least once within each backup level. For instance, it was necessary to have at least 4 and 7 ambulances, respectively, in the first and the second small example to avoid infeasibility. We do not recommend to relax the constraints (8.1b), although this can guarantee feasibility. Some demand zones might not be covered even within the first backup level after relaxing these constraints which is not practical.

### 8.4.2 Model Evaluation

We develop a discrete-event simulation to evaluate the performance of the SSDD solution to the large instance based on the real dataset. We generate emergency calls to the Mecklenburg EMS agency during 2 weeks, i.e., 28 12-h shifts. We consider 18, 19, and 20 ambulances and sample 450 scenarios. This fleet size is the common range of available ambulances in Mecklenburg county. All reported instances are solved optimally within 4 h. We could not increase the scenario size more than 450 due to the computational intractability of solving the extensive form of the model. For instance, when the number of scenarios is 500 and the fleet size is 20, the best reported gap by CPLEX is 27.5% after 4 h run time. A tailored decomposition algorithm might be developed to solve large scale instances rather than solving the extensive form of the model. This is beyond the scope of this chapter, and we refer the reader to Enayati et al. [18] and Lei et al. [35] for more information on decomposition methods in a similar application domain. We assume that ambulances are initially located at the sites proposed by the SSDD model and they return to their home sites after service completion. Ambulances can be dispatched to new calls before returning to their home sites. Furthermore, if more than one ambulance is available for a call, the closest one is dispatched. We compute actual road network distance for all ambulance trips using the Matlog toolbox in MATLAB [29].

It might not be possible to apply optimal dispatching decisions in real life due to legal regulations that mandate sending the closest available ambulance to an emergency call. Hence, we study the isolated effect of ambulance location and dispatching decisions on the performance of an EMS system by considering two additional baseline policies. The *baseline policy-1* merely uses the ambulance sites in the solution to the SSDD model. In this baseline policy, ambulances are dispatched according to the myopic policy that sends the closest available ambulance to an emergency call. In *baseline policy-2* ambulances are initially located based on the Maximum Covering Location Problem (MCLP) [13]. We use the historical proportional demand corresponding to each demand zone as the weight multiplier in the objective function of the MCLP model. Dispatching policy in the *baseline policy-2* also sends the closest available ambulance. We report two performance measures along with the computational time to solve the extensive form of the SSDD model in Table 8.2:

- MRT (min): mean response time to calls during each shift.
- Workload (%): average percentage of time that an ambulance is busy with service and travels (dispatching and transportation to hospital) during each shift.

Table 8.2 95% confidence intervals of mean response time and workload measures for 2-week simulation of an EMS system

| | SRDP | Baseline-1 | Baseline-2 |
|---|---|---|---|
| $\kappa = 18$ | | | |
| MRT (min) | [9.9, 13.7] | [11.3, 15.1] | [11.9, 15.8] |
| Workload (%) | [54.8, 59.6] | [58.9, 62.6] | [59.5, 63.2] |
| Time (s) | 7982 | – | – |
| $\kappa = 19$ | | | |
| MRT (min) | [9.2, 13.3] | [10.9, 14.9] | [11.2, 15.1] |
| Workload (%) | [50.2, 52.8] | [53.3, 56.5] | [53.6, 56.2] |
| Time (s) | 8235 | – | – |
| $\kappa = 20$ | | | |
| MRT (min) | [8.8, 12.6] | [10.5, 14.0] | [10.6, 14.9] |
| Workload (%) | [43.2, 45.7] | [47.1, 49.3] | [47.8, 49.7] |
| Time (s) | 9998 | – | – |

The reported time indicates the computational time to obtain the solution of solving the extensive form of the SSDD model with 450 scenarios by CPLEX

We make the following observations from Table 8.2:

- Full implementation of the SSDD model always outperforms both baseline policies in terms of mean response time. Comparing to the *baseline policy-2* in which alternative location and dispatching policies are applied, the full implementation of the SSDD model improves the mean response time by 2 min on average. We also observe that solely implementing the SSDD location solution with an alternative dispatching policy, as in *baseline policy-1*, does not improve the mean response time as much.
- Full implementation of the SSDD model always imposes less workload on ambulances by about 4% on average as compared with both baseline policies. The workload of ambulances in *baseline policy-1* and *baseline policy-2* is statistically similar.

Finally, we calculate the value of stochastic solution that measures the potential benefit of solving the stochastic program over solving a deterministic program in which random parameters are replaced with their expectations [9]. That is, we generate a single scenario in which all random parameters (e.g., $d_i^s$) are estimated to the average of all 450 generated scenarios. We then run the SSDD for this scenario and evaluate the output using simulation. The value of stochastic solution in our set of experiments is reported 18% and 29% based on the MRT and the workload measures, respectively.

## 8.5 Conclusion

In this chapter, we formulate a two-stage stochastic program to address the service district design problem. This model presents two outputs: (1) available ambulances are located at the potential sites in the service area, and (2) response districts for each ambulance are designed at different backup levels (i.e., each ambulance is assigned to serve a set of demand zones within a certain response threshold). We maximize the expected proportion of covered calls while restricting the workload of ambulances. The proposed model assumes that each ambulance has identical and independent busy fraction probability.

We illustrate the solution on two small examples. We then evaluate the implementation aspects of the proposed model via a discrete-event simulation by comparing the model with two baseline policies to study the isolated impact of location and dispatching decisions resulting from the proposed model. The two baseline policies are designed due to the mandatory regulation in most EMS systems that obligates to send the closest available ambulance to an emergency call. Our computational results show a significant improvement in mean response time by 2 min on average when implementing the proposed model. We also observe that average workload of ambulances is statistically less by 4% on average when the model is fully implemented.

Future studies may formulate a model that endogenously estimates the busy fraction of ambulances as a decision variable for the service district design problem. The results of such a formulation can be compared with the proposed model in this chapter to study the impact of incorporating the ambulance busy fractions exogenously versus endogenously on the quality of solutions. An endogenous formulation could also alleviate the possibility of infeasible solutions. Furthermore, future studies may develop a decomposition algorithm for the proposed two-stage stochastic program to deal with computational intractability due to large number of scenarios.

**Acknowledgements** The authors would like to thank Professor Cem Saydam from the Belk College of Business at the University of North Carolina, Charlotte, for sharing the large dataset used in this research and for his continued support.

# References

1. Alanis, R., Ingolfsson, A., Kolfal, B.: A Markov chain model for an EMS system with repositioning. Prod. Oper. Manag. **22**(1), 216–231 (2013)
2. Andersson, T., Varbrand, P.: Decision support tools for ambulance dispatch and relocation. J. Oper. Res. Soc. **58**(2), 195–201 (2007)
3. Ansari, S., McLay, L., Mayorga, M.: A maximum expected covering problem for district design. Transp. Sci. **51**(1), 376–390 (2017)
4. Aringhieri, R., Bruni, M., Khodaparasti, S., van Essen, J.: Emergency medical services and beyond: addressing new challenges through a wide literature review. Comput. Oper. Res. **78**, 349–368 (2017)
5. Başar, A., Çatay, B., Ünlüyurt, T.: A taxonomy for emergency service station location problem. Optim. Lett. **6**(6), 1147–1160 (2012)
6. Batta, R., Dolan, J.M., Krishnamurthy, N.N.: The maximal expected covering location problem: revisited. Transp. Sci. **23**(4), 277–287 (1989)
7. Bélanger, V., Ruiz, A., Soriano, P.: Recent advances in emergency medical services management. Technical Report FSA-2015-006. Faculté des Sciences de l'Administration, Université Laval, Quebec City (2015)
8. Beraldi, P., Bruni, M.: A probabilistic model applied to emergency service vehicle location. Eur. J. Oper. Res. **196**(1), 323–331 (2009)
9. Birge, J.R.: The value of the stochastic solution in stochastic linear programs with fixed recourse. Math. Program. **24**(1), 314–325 (1982)
10. Brotcorne, L., Laporte, G., Semet, F.: Ambulance location and relocation models. Eur. J. Oper. Res. **147**(3), 451–463 (2003)
11. Budge, S., Ingolfsson, A., Erkut, E.: Approximating vehicle dispatch probabilities for emergency service systems with location-specific service times and multiple units per location. Oper. Res. **57**(1), 251–255 (2009)
12. Caro, F., Shirabe, T., Guignard, M., Weintraub, A.: School redistricting: embedding GIS tools with integer programming. J. Oper. Res. Soc. **55**(8), 836–849 (2004)
13. Church, R., ReVelle, C.: The maximal covering location problem. Pap. Reg. Sci. Assoc. **32**(1), 101–118 (1974)
14. Daskin, M.: A maximal expected covering location model: formulation, properties and heuristic solution. Transp. Sci. **17**(1), 48–70 (1983)
15. Daskin, M.S., Stern, E.H.: A hierarchical objective set covering model for emergency medical service vehicle deployment. Transp. Sci. **15**(2), 137–152 (1981)

16. Duque, J.C., Church, R.L., Middleton, R.S.: The p-regions problem. Geogr. Anal. **43**(1), 104–126 (2011)
17. Enayati, S., Mayorga, M.E., Rajagopalan, H.K., Saydam, C.: Real-time ambulance redeployment approach to improve service coverage with fair and restricted workload for EMS providers. Omega **79**, 67–80 (2018)
18. Enayati, S., Özaltın, O.Y., Mayorga, M.E., Saydam, C.: Ambulance redeployment and dispatching under uncertainty with personnel workload limitations. IISE Trans. **50**(9), 777–788 (2018)
19. Erkut, E., Ingolfsson, A., Sim, T., Erdogan, G.: Computational comparison of five maximal covering models for locating ambulances. Geogr. Anal. **41**(1), 43–65 (2009)
20. Gendreau, M., Laporte, G., Semet, F.: The maximal expected coverage relocation problem for emergency vehicles. J. Oper. Res. Soc. **57**(1), 22–28 (2006)
21. Geroliminis, N., Karlaftis, M., Skabardonis, A.: A spatial queuing model for the emergency vehicle districting and location problem. Transp. Res. B Methodol. **43**(7), 798–811 (2009)
22. Goldberg, J.B.: Operations research models for the deployment of emergency services vehicles. EMS Manag. J. **1**(1), 20–39 (2004)
23. Grannan, B., Bastian, N., McLay, L.: A maximum expected covering problem for locating and dispatching two classes of military medical evacuation air assets. Optim. Lett. **9**(8), 1511–1531 (2015)
24. Haugland, D., Ho, S.C., Laporte, G.: Designing delivery districts for the vehicle routing problem with stochastic demands. Eur. J. Oper. Res. **180**(3), 997–1010 (2007)
25. Henderson, S.G., Mason, A.J.: Ambulance service planning: simulation and data visualisation. In: Brandeau, M.L., Sainfort, F., Pierskalla, W.P. (eds.) Operations Research and Health Care, International Series in Operations Research & Management Science, vol. 70, pp. 77–102. Springer, Boston (2005)
26. Iannoni, A.P., Morabito, R., Saydam, C.: An optimization approach for ambulance location and the districting of the response segments on highways. Eur. J. Oper. Res. **195**(2), 528–542 (2009)
27. Ingolfsson, A.: EMS planning and management. In: Zaric, G. (ed.) Operations Research and Health Care Policy, International Series in Operations Research & Management Science, vol. 190, pp. 105–128. Springer, New York (2013)
28. Kalcsics, J.: Districting problems. In: Laporte, G., Nickel, S., Saldanha da Gama, F. (eds.) Location Science, pp. 595–622. Springer, Cham (2015)
29. Kay, M.G.: Matlog: Logistics Engineering Matlab Toolbox (2017). Available at http://www4.ncsu.edu/~kay/matlog/. Last Accessed 10 Aug 2017
30. Kong, Y., Zhu, Y., Wang, Y.: A center-based modeling approach to solve the districting problem. Int. J. Geogr. Inf. Sci. **33**(2), 368–384 (2019)
31. Larsen, M.P., Eisenberg, M.S., Cummins, R.O., Hallstrom, A.P.: Predicting survival from out-of-hospital cardiac arrest: a graphic model. Ann. Emerg. Med. **22**(11), 1652–1658 (1993)
32. Larson, R.: A hypercube queuing model for facility location and redistricting in urban emergency services. Comput.Oper. Res. **1**(1), 67–95 (1974)
33. Lee, S.: The role of preparedness in ambulance dispatching. J. Oper. Res. Soc. **62**(10), 1888–1897 (2011)
34. Lei, H., Laporte, G., Guo, B.: Districting for routing with stochastic customers. EURO J. Transp. Logist. **1**(1–2), 67–85 (2012)
35. Lei, C., Lin, W.H., Miao, L.: A stochastic emergency vehicle redeployment model for an effective response to traffic incidents. IEEE Trans. Intell. Transp. Syst. **16**(2), 898–909 (2015)
36. Li, Z., Wang, R.S., Wang, Y.: A quadratic programming model for political districting problem. In: Du, D.Z., Zhang, X.S. (eds.) Proceedings of the First International Symposium on Optimization and System Biology (OSB'07), Beijing, pp. 427–435 (2007)
37. Li, X., Zhao, Z., Zhu, X., Wyatt, T.: Covering models and optimization techniques for emergency response facility location and planning: a review. Math. Meth. Oper. Res. **74**(3), 281–310 (2011)

38. Liu, Y., Yuan, Y., Li, Y., Pang, H.: A chance constrained programming model for reliable emergency vehicles relocation problem. Procedia Soc. Behav. Sci. **96**, 671–682 (2013)
39. Marianov, V., ReVelle, C.: The queuing probabilistic location set covering problem and some extensions. Socio Econ. Plan. Sci. **28**(3), 167–178 (1994)
40. Marianov, V., ReVelle, C.: The queueing maximal availability location problem: a model for the siting of emergency vehicles. Eur. J. Oper. Res. **93**(1), 110–120 (1996)
41. Maxwell, M., Restrepo, M., Henderson, S., Topaloglu, H.: Approximate dynamic programming for ambulance redeployment. INFORMS J. Comput. **22**(2), 266–281 (2010)
42. Maxwell, M., Ni, E., Tong, C., Handerson, S., Topaloglu, H., Hunter, S.: A bound on the performance of an optimal ambulance redeployment policy. Oper. Res. **62**(5), 1014–1027 (2014)
43. Mayorga, M., Bandara, D., McLay, L.: Districting and dispatching policies to improve the efficiency of emergency medical service (EMS) systems. IIE Trans. Healthc. Syst. Eng. **3**(1), 39–56. (2013)
44. McLay, L., Mayorga, M.: Evaluating the impact of performance goals on dispatching decisions in emergency medical service. IIE Trans. Healthc. Syst. Eng. **1**(3), 185–196 (2011)
45. McLay, L., Mayorga, M.: A dispatching model for server-to-customer systems that balances efficiency and equity. Manuf. Serv. Oper. Manag. **15**(2), 205–220 (2012)
46. McLay, L.A., Mayorga, M.E.: A model for optimally dispatching ambulances to emergency calls with classification errors in patient priorities. IIE Trans. **45**(1), 1–24 (2013)
47. Mehrotra, A., Johnson, E.L., Nemhauser, G.L.: An optimization based heuristic for political districting. Manag. Sci. **44**(8), 1100–1114 (1998)
48. Moore, G., ReVelle, C.: The hierarchical service location problem. Manag. Sci. **28**(7), 775–780 (1982)
49. Naoum-Sawaya, J., Elhedhli, S.: A stochastic optimization model for real-time ambulance redeployment. Comput. Oper. Res. **40**(8), 1972–1978 (2013)
50. Nemoto, T., Hotta, K.: Modelling and solution of the problem of optimal electoral districting. Commun. Soc. Jap. **48**, 300–306 (2003)
51. Pell, J.P., Sirel, J.M., Marsden, A.K., Ford, I., Cobbe, S.M.: Effect of reducing ambulance response times on deaths from out of hospital cardiac arrest: cohort study. Br. Med. J. **322**(7299), 1385–1388 (2001)
52. Puppe, C., Tasnádi, A.: A computational approach to unbiased districting. Math. Comput. Model. **48**(9–10), 1455–1460 (2008)
53. Rajagopalan, H.K., Saydam, C., Xiao, J.: A multiperiod set covering location model for dynamic redeployment of ambulances. Comput. Oper. Res. **35**(3), 814–826 (2008)
54. ReVelle, C.S., Hogan, K.: The maximum availability location problem. Transp. Sci. **23**(3), 192–200 (1989)
55. Ricca, F., Scozzari, A., Simeone, B.: Political districting: From classical models to recent approaches. Ann. Oper. Res. **204**(1), 271–299 (2013)
56. Riverside County EMS Agency: Riverside County EMS system strategic plan 2014–2019 – ver. 9-4-14. Report (2014). Available at http://remsa.us/documents/plans/140923FINALEMSSystemStratPlan.pdf
57. Salazar-Aguilar, M.A., Ríos-Mercado, R.Z., González-Velarde, J.L.: A bi-objective programming model for designing compact and balanced territories in commercial districting. Transp. Res. C Emerg. Technol. **19**(5), 885–895 (2011)
58. Saydam, C., Rajagopalan, H., Sharer, E., Lawrimore-Belanger, K.: The dynamic redeployment coverage location model. Health Syst. **2**(2), 103–119 (2013)
59. Schmid, V.: Solving the dynamic ambulance relocation and dispatching problem using approximate dynamic programming. Eur. J. Oper. Res. **219**(3), 611–621 (2012)
60. Schmid, V., Doerner, K.: Ambulance location and relocation problems with time-dependent travel times. Eur. J. Oper. Res. **207**(3), 1293–1303 (2010)
61. Sherali, H.D., Smith, J.C.: Improving discrete model representations via symmetry considerations. Manag. Sci. **47**(10), 1396–1407 (2001)

62. Sudtachat, K., Mayorga, M., McLay, L.: Recommendations for dispatching emergency vehicles under multitiered response via simulation. Int. Trans. Oper. Res. **21**(4), 581–617 (2014)
63. Sudtachat, K., Mayorga, M., McLay, L.: A nested-compliance table policy for emergency medical service systems under relocation. Omega **58**, 154–168 (2016)
64. Teixeira, J.C., Antunes, A.P.: A hierarchical location model for public facility planning. Eur. J. Oper. Res. **185**(1), 92–104 (2008)
65. Toregas, C., Swain, R., ReVelle, C., Bergman, L.: The location of emergency service facilities. Oper. Res. **19**(6), 1363–1373 (1971)
66. Toro-Díaz, H., Mayorga, M.E., Chanta, S., McLay, L.A.: Joint location and dispatching decisions for emergency medical services. Comput. Ind. Eng. **64**(4), 917–928 (2013)

# Part III
# Applications and Case Studies

# Chapter 9
# Spatial Optimization Problem for Locating Polling Facilities and Stations and Policy Implications

**Hyun Kim and Kamyoung Kim**

**Abstract** Voting is a critical political activity and gives voters the opportunity and right to express their opinion in modern democratic society. Ways to increase voter turnout have been widely explored, but, the optimization approach is recognized by many scholars as the best way to assess the efficiency of the current system and draw policy implications. This research highlights the necessity for a spatial optimization approach in determining the location of polling facilities and polling stations tailored to the regulations of the voting process of South Korea. The effects of distance and preference, such as that based on pre-knowledge of or experience with existing facilities, are prescribed as the function 'utility cost' in formulating a spatial optimization model, named the *Capacitated p-Median Problem with Multiple Stations in the Same Facility (CPMP-M)*. In a case study of an area with several precincts in Seoul, South Korea, our numerical results based on preference factors demonstrate the need to relocate the existing polling facilities, merge certain precincts, and adjust existing boundaries of precincts to enhance the efficiency of administration of the voting process.

## 9.1 Introduction

Voting is the most fundamental method of political participation in modern democracy. The most important question regarding voting behavior is as follows: What are the critical factors that motivate eligible voters to participate in voting, and how can the rate of participation be improved [11]? This question is critical because

H. Kim (✉)
Department of Geography, University of Tennessee at Knoxville, Knoxville, TN, USA
e-mail: hkim56@utk.edu

K. Kim
Department of Geography Education, Kyungpook National University, Buk-gu, Daegu, South Korea
e-mail: kamyoungkim@knu.ac.kr

© Springer Nature Switzerland AG 2020
R. Z. Ríos-Mercado (ed.), *Optimal Districting and Territory Design*, International Series in Operations Research & Management Science 284,
https://doi.org/10.1007/978-3-030-34312-5_9

different political turnouts can directly affect election results, and depending on voter turnout, the outcome of an election may not sufficiently reflect the preferences of the community [42]. Though there are many reasons that eligible voters do not participate in voting, several theories have explained the motivation in terms of economic perspectives. For example, Downs [6] explained the motivation of voters using the rational choice theory. This theory employed a dichotomy of economic behavior: cost and benefit. Voters engage in the voting process only if the benefits of voting are greater than the costs resulting from voting. Although Downs's theory provided a general frame to understand voting behaviors, it raised the question of which elements are crucial factors in terms of the costs of voting. One of the costs explored was the concept of accessibility of polling stations to eligible voters within a precinct [7, 13, 32].

Ease of access to a voting location (e.g., ballot box, station, facility) can be translated into the function of cost, which is well quantified with a geographic factor, "distance." Brady and McNulty [3] highlighted the importance of distance to understanding the behavior of voters through their study of the relationship between locations of polling places and costs. The cost of a longer distance of travel from a voter's residence to a polling station leads to an increased utility cost. If the cost is high enough that eligible voters have a benefit smaller than the cost, then they are not likely to participate in the voting process in a practical sense [2]. For example, in the case of Seoul, South Korea, the turnout rates of voters in the 20th National Assembly election of 2016 ranged from 26.9 to 72.3% among 2258 precincts, indicating that these rates could be highly associated with the cost factors of distance from the eligible voters in a precinct to the designated voting facilities [36]. As pointed out by Kim and Kim [20, 21], the decreased rate of voting in South Korea is recognized as a social problem, requiring an assessment of the current placement of polling stations [22, 23]. To solve this problem, the spatial optimization approach is considered to assess current system efficiency in terms of administration and improve accessibility of voters by delineating precincts and selecting polling facilities and placement of polling stations within the given regulations for better political participation [2, 3, 27].

The goal of this paper is to highlight the necessity for a spatial optimization approach in determining the location of polling facilities and polling stations tailored to the regulations of the voting process of South Korea. The effects of distance and preference based on things such as pre-knowledge of or experience with existing facilities, are prescribed as the function "utility cost" in formulating a spatial optimization model, named the *Capacitated p-Median Problem with Multiple Stations in the Same Facility (CPMP-M)*. The structure of the model is formulated based upon the traditional $p$-median problem; however, the restrictions imposed by the administrative requirements were specified as a set of constraints. More important, voters' preferences in selecting polling facilities or stations were added to the distance effect for traveling cost, which is widely used in the majority of location problems. In addition, demand areas were created by re-aggregated household data, which takes into account the unique characteristics of housing systems in Seoul. Our study explores the relationships between preference factors and the response of the models in deciding the location of polling facilities and the allocation of demand areas. Finally, several policy implications are drawn from the results.

## 9.2   Background

Many studies have examined the effect of distance on voter turnout. Previous research on voting behaviors has focused on the intangible benefits of voting activity, such as fulfilling one's responsibility as a citizen at an individual level, in seeking to understand voting participation. The aspects of cost, however, were largely ignored because of the difficulty of quantifying costs. The factors of costs are usually based on intangible motivation. Distance is an important metric that affects an eligible voter's decision-making process in regard to participation, because participation incurs the direct cost of reaching the polling facility or stations during limited voting hours. However, in the context of the USA, transportation costs have been largely ignored in understanding political participation [1] because there are few effective instruments to measure the transportation costs of voters at an individual level, and the effect of distance has been generally approximated based on the aggregated level of voters in specified spatial units [13].

A handful of studies found that distance had a significant effect on voting turnout (see [2, 7, 13]). A lack of motivation to vote has also been recognized as a factor. However, it is known that improving the convenience of voting by providing better access to polling places could produce higher turnouts [11]. The underlying assumption of the role of distance on voting turnout is that the greater the distance from residence to a polling place, the more the opportunity cost of voting negatively affects voting participation. Furthermore, recent research highlighted that not only the cost required to access voting places but other intangible costs, such as the searching cost to find polling locations, can be added to understand voting behaviors [3]. Recent empirical research claimed that the effect of distance as a cost can be captured much more clearly with the help of Geographic Information Systems, which make more detailed information available (e.g., the locations of polling stations and options to select among candidate sites), confirming that the distance from voters to polling facilities negatively affects voting turnouts [7, 19].

The problem of placing voting facilities and stations (i.e., voting system) is more critically associated with the effectiveness of administration, while an electoral system focuses more on the issue of delineation of electoral districts, which is grounded in political equity [26, 44]. The voting system requires managing the entire process of voting, which should provide better convenience for eligible voters to encourage the participation in voting, resulting in higher turnout rates [17]. The convenience of voting is directly related to the ease of access to voting facilities, often called "accessibility" [26, 43]. Though the importance of the location problem of polling facilities has been stressed, there is a lack of literature addressing it [8, 37, 39].

Two perspectives justify the need to improve accessibility of eligible voters to polling places. First, constraining voting activity due to far distance violates the principle of suffrage in a modern democracy. To encourage eligible voters to participate in politics, equal opportunity should be ensured and any physical hindrance should be eliminated [2]. In this perspective, improving accessibility can result in more equal opportunity, which is anticipated to increase participation. Second, from a practical perspective, enhancing accessibility minimizes the operational cost of administration. The establishment of precincts and polling stations is a critical election administration process and entails a significant expense for the public office that runs the event. Thus, both increased turnouts and reduced operation costs should be reasonably sought [40]. Adjusting the location of polling facilities and stations is key to producing a better turnout rate, though it might not be a perfect policy instrument. As empirical examples, Brady and McNulty [3] reported that reducing the number of polling stations by consolidating precincts cuts down election costs in a case study of Los Angeles County in the USA for the 2002 and 2003 election. The research argued that a tiny increase in the cost of traveling to a voting place may reduce little motivation to vote, because distance could be a crucial factor for voters. Bhatti [2] also highlighted the importance of location selection of facilities to reduce traveling costs with a case study of Denmark, while Konishi et al. [27] and Murata and Konishi [33] discussed the need for modeling to reduce costs by using allocation models that focus on minimizing the distance from voters to stations. In detail, Konishi et al. [27] employed a heuristic algorithm to find a solution that would reduce both the distance of voters to facilities and the cost of those facilities. Cantoni [4] developed an allocation problem using mixed integer programming (MIP), in which the objective function was to minimize the cost of moving voters to facilities while balancing the demand and capacity of the facilities. In the context of political voting systems in South Korea, however, only a few studies have analyzed the effect of distance and assessed the location of electoral districts and voting places [20, 21, 31]. The studies raised the need for a mathematical modeling approach which reflects not only the effect of distance but also the stipulations of the official voting and election codes of South Korea. Note that the requirements have changed for each voting event because of the dynamic change in registered voters as well as the boundaries of spatial units [14, 15]. Most research, however, has focused on an analysis of voting results after elections. Virtually none of the research has sought to provide answers regarding the optimal arrangement of voting places in light of eligible voters (i.e., demands) or suggested an adjustment of precincts which can improve the efficiency of administration and voting turnouts.

The remainder of this paper is organized as follows: Sect. 9.3 explores the core modeling issues in formulating a location model using the $p$-median problem when the model takes account of voting behaviors and requirements in the election codes of South Korea. The study area is carefully selected in terms of size in order to solve the various instances using the prescribed model in the environment of commercial optimization software (CPLEX). Section 9.4 presents the mathematical formulation of the model, followed by the computational results in Sect. 9.5, focusing on the variation of optimal arrangements with the different settings of utility costs. The concluding remarks section provides several policy implications from the results.

## 9.3 Modeling Issues and Data

Generally, five issues are raised when a spatial optimization approach is considered in voting place location models: (1) representation of demand, (2) traveling cost, (3) voters' preference for polling facilities, (4) spatial scope of modeling, and (5) capacity of polling stations. First, the representation issue has long been recognized as a crucial issue in location modeling, as the space should be discretized regarding the level of aggregation from different spatial units. Note that the quality of solutions of optimization models varies with the spatial units used [34, 35]. In theory, using the smallest spatial unit possible, such as the individual level, could be the ideal way to reduce the propagation of errors of uncertainty, potential problems of *gerrymandering*, and the effects of the modifiable areal unit problem (MAUP), which has been a critical issue in designing voting systems with different districting schemes [16]. However, the spatial units for administrative purposes are usually delineated at a certain aggregated level with an extent of space. According to the specified aggregated level, the data at the individual level should be aggregated to represent the properties of variables at the aggregated level. Often, existing irregular spatial units must be treated as regular and smaller units through the tessellation process or statistical clustering methods to replace the distribution of existing values at the original units with estimated values of the new units [28, 30, 45].

When disaggregated data such as the individual level are aggregated as the input for spatial optimization problems, two issues are raised. First, the size of input data is of concern in terms of the complexity of the location problem, because it affects the capability of mathematical models to solve instances with optimality. Obviously, the increased size of input data entails a significant computational burden, often requiring the development of a heuristic algorithm which trades off between the quality of solutions and solution times. This is particularly true for the $p$-median problem and district delineation problems [18, 24, 25]. If it is inevitable to aggregate the input data, the method of aggregation then becomes a following issue. In our case study, considering the uniqueness of the housing system in South Korea, where most residential areas are highly dense due to a number of apartment complexes and multi-unit dwelling housing systems, representing demand areas at an individual level (i.e., each person) is not recommended, but the household level is recognized as the proper spatial units for analysis (see [29, 30]). We generate demand areas at a household level using registered voters' addresses. The aggregation is made using a sub-unit of households, which is a unique system of the address system under the administrative unit of *Dong*. The aggregated areas are represented as a regularly sized spatial point unit $a_i$ using Eq. (9.1).

$$E(a_i) = \frac{S_i}{T_j} \times R_j \qquad (9.1)$$

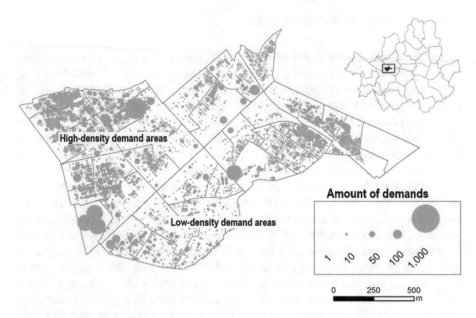

**Fig. 9.1** Spatial distribution of demands in *Seogyo-Dong, Mapo-Gu* in Seoul

where $S_i$ is the number of households in a sub-unit in a precinct $j$, $T_j$ is the total number of households in a precinct $j$, and $R_j$ is the total number of registered voters in a precinct $i$.

Figure 9.1 illustrates the distribution of demands within the study area of *Seogyo-Dong, Mapo-Gu* in Seoul using Eq. (9.1). Currently, the area consists of seven precincts with 2497 aggregated demand points from 21,348 eligible voters.

Second, travel cost is well recognized as a fundamental element in location problems, because both the decisions about where to locate facilities and how to assign demands to the selected facilities are greatly affected by physical hindrance, the friction of distance. Thus, the traveling cost is represented as a function of a type of distance. In general, either physical or time distance is used as a proxy for the traveling cost. In political districting problems, distance is an important determinant for eligible voters in selecting the best polling place among candidate sites. Moreover, recent empirical research demonstrated that traveling cost is a principal factor to voting turnout [37]. The closer voters are to the polling facility, the greater the probability they will select the facility to complete their voting activity. The distance-based arrangement of location models suggests that changing the location of voting facilities, installing new voting places, or consolidating or adjusting existing precincts is key to enhancing the objective function which is to minimize the traveling cost.

Third, voters' preference for polling facilities is an important factor in their decision to select which station they visit. From a behavioral or psychological perspective, it is recognized that preference helps voters reduce the searching cost

[7]. The factor, however, has rarely been addressed in the literature since the initial work applying to political districting facility location problems by Hanjoul and Peeters [12]. Preference for specific facilities is associated with voters' familiarity with existing or previous facilities they have used as voting places before as well as with their experience when they participated in elections. According to the election results for 2016 in South Korea, voters' preference for the use of public facilities (e.g., schools, community centers) was much higher (58.8%) than for private facilities (3.2% for banks, daycare centers, and wedding halls) because public facilities are usually more visible and accessible daily than private facilities. Historically, the more a facility is used as a voting place for an election, the more accustomed voters will be to the facility for the next election. As result, increased familiarity with the facilities affects the voters' preferences when selecting voting facilities and stations. The issue in practice is how to quantify the preferences. In our study, we scaled preference as a number between 1 and 10 based on the percentage of usage of facilities in National Assembly election data [36]. Figure 9.2 presents the distribution of polling facilities (a) and their preference scores (b).

Fourth, the spatial scope of modeling, especially when assigning demand areas to facilities (or polling stations), is an important issue because the spatial scope is delineated as the regulation for a voting event. In South Korea's voting system, the selection of polling facilities as well as the installation of polling stations should comply with the requirements of the Public Official Election Act (POEA), which was enacted by the Statutes of the Republic of Korea. Often, the rules are revised before an election, and regulations related to spatial scope should be adhered to. According to the POEA 2016, a set of precincts is delineated within a spatial administrative unit of "county (*Gu*)." A *Gu* may have several precincts for a voting event. In the POEA 2016, the term *polling facility* refers to a physical building in a precinct where voters should go for the voting activity. The *polling station*, often called the *ballot box*, should be placed in a polling facility. Note that a polling facility may have multiple polling stations. However, the number of facilities and polling stations is determined by the central and local governments according to regulations or the requirements described in the POEA [41]. In essence, the location of polling facilities and stations should satisfy three important requirements of the POEA:

- One basic administrative spatial unit (*Gu*) consists of several precincts to improve the efficiency of voting process (Article 31).
- A precinct may have more than one polling station in a polling facility (Article 31).
- If there is no suitable facility to install the polling station in the area, it may be installed in adjacent precinct, and polling stations may be located outside of the County (Article 147-2).

Figure 9.3 illustrates the process of locating polling facilities and polling stations and their allocation from demands in precincts. Among candidate polling facilities in a precinct, the best location for the facility is determined based on the principal factors, such as travel costs and the degree of preference for facilities. Other factors

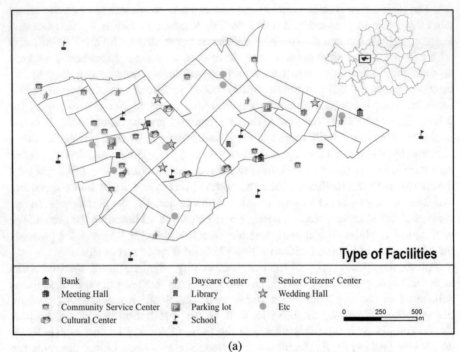

(a)

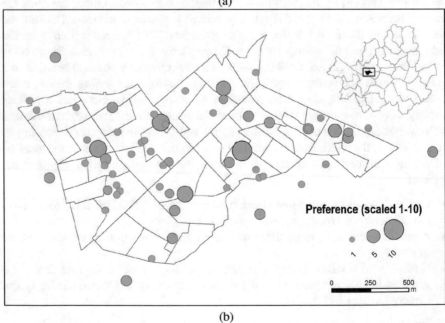

(b)

**Fig. 9.2** Types of polling facilities (**a**) and preference of facilities (**b**)

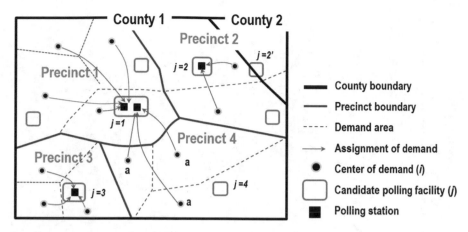

**Fig. 9.3** Location of polling facilities and polling stations, and allocation of demand

being equal, the location of the facility in a precinct is purely determined by the total weighted sum of distance from demands to the facility, which generates the least traveling cost of voters or demand areas. For example, the arrangement of precinct 3 falls into this case, with three demand areas assigned to polling station 3. Given the second regulation, the polling station for the demand areas "*a*" of precinct 4 can be placed in facility 1 in precinct 1 if facility 4 cannot serve the demand areas for some reason. Facility 2 in precinct 2 shows that a polling facility for a precinct can be located in a different county if there is no suitable facility within a corresponding precinct.

Fifth, the capacity of a polling station is associated with the voting hours and the number of personnel at the polling facility, which limits the number of voters allocated to that polling facility. Figure 9.4 presents the relationship between the number of voters and turnout rate from the result of the 20th National Assembly election in 2016. Notice that the number of voters and the turnout rate are related to each other, indicating that if the capacity of a polling facility is too small or too large, turnout rate is diminished. Based on the result, the ideal range of capacity is determined to be [2192 − 4297 $persons/facility$].

## 9.4   Model Formulation

The *p*-median problem is a classical location problem of locating *p* facilities to minimize the weighted average cost between demands and the selected facilities [5]. The problem is a class of location-allocation, which determines the location of facilities while at the same time considering the allocation of demands to facilities. The problem can be extended by plugging in capacity of facility [9, 10], named the capacitated *p*-median problem. This study presents the Capacitated *p*-Median

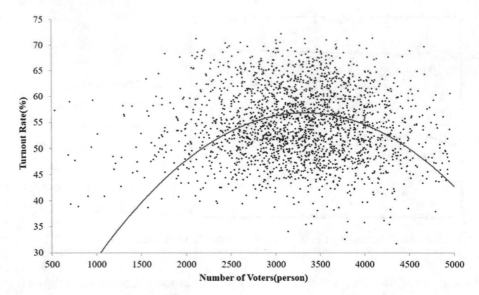

**Fig. 9.4** Relationship between the number of voters and turnout rate of the 20th national assembly election, 2016 (Seoul, South Korea)

Problem with Multiple Stations in the Same Facility (CPMP-M), which is an extended version of the $p$-median problem. The problem locates $p$ capacitated facilities in a given space, with a facility able to reside in multiple stations, if necessary, to minimize the total utility costs from demands (i.e., unit of aggregated voters) to be assigned to polling facilities and stations. A new term, *"utility cost,"* here is defined as a function with a combination of factors such as traveling cost (i.e., physical distance) and facility preference. As introduced by Ghiani et al. [9] in their models, a unique situation in which multiple facilities are placed in a single site was reflected in the model formulation. The CPMP-M includes a polling station location problem, in which several identical facilities can be opened to accommodate multiple polling stations in the same facility. The CPMP-M is formulated as a mixed integer linear program (MILP), as follows:

*Indices and Parameters*

| | |
|---|---|
| $i, k$ | = index of demand points, $i, k \in I$, |
| $j$ | = index of facilities, $j \in J$, |
| $a_i$ | = the population of demand point $i$, |
| $d_{ij}$ | = distance between demand $i$ and facility $j$, |
| $f_j$ | = preference of facility $j$ defined in terms of type and history, |
| $\alpha$ | = scaling factor, |
| $\beta$ | = preference impact factor, |
| $p$ | = number of facilities to be cited, |
| $CLx_j$ | = lower bound of workload at facility $j$, |
| $CUx_j$ | = upper bound of workload at facility $j$. |

*Decision Variables*

$x_j = $ number of (multiple) polling stations placed at facility $j$

$$y_{ij} = \begin{cases} 1, & \text{if demand } i \text{ is assigned to facility } j \\ 0, & \text{otherwise.} \end{cases}$$

*Formulation*

$$\text{minimize} \quad \Omega = \alpha \sum_i \sum_j a_i \frac{d_{ij}}{f_j^\beta} y_{ij} \tag{9.2a}$$

$$\text{subject to} \quad \sum_j x_j = p \tag{9.2b}$$

$$\sum_j y_{ij} = 1 \qquad\qquad i \in I \tag{9.2c}$$

$$y_{ij} \leq x_j \qquad\qquad i \in I, j \in J \tag{9.2d}$$

$$x_j \geq 0 \qquad\qquad j \in J \tag{9.2e}$$

$$\sum_i a_i y_{ij} \leq CU x_j \qquad\qquad j \in J \tag{9.2f}$$

$$\sum_i a_i y_{ij} > Cl_. x_j \qquad\qquad j \subset J \tag{9.2g}$$

$$y_{ij} \in \{0, 1\} \qquad\qquad i \in I, j \in J \tag{9.2h}$$

$$x_j \in \{0, 1, 2, \ldots\} \qquad\qquad j \in J \tag{9.2i}$$

The objective function minimizes the sum of the total utility cost $(d_{ij}/f_j^\beta)$ from demands $i$ to facilities $j$. Here, the utility cost consists of the factors of traveling cost (distance $d_{ij}$) and voters' preference function $f_j$ for facility $j$. Notice that the preference function $f_j$ is treated as the denominator to distance effect. If distance effect is equal for all facilities $j$ from demand $i$, then the decision of demand $i$ will be made based on the preference factor of the facility. In other words, if voters' familiarity and existing experience with the facility $j$ outweigh the distance effect, then voters will accept the extra traveling cost to go farther to vote, but this decision may imply that voters save the search cost of finding a nearby polling facility. Notice that the exponent $\beta$, which is added as a parameter of $f_j$, reflects the degree of power of the preference factors. By changing the value with a range of $[0, 3]$, the effect of preference will be examined. The results will help in understanding the behaviors of voters, depending on the friction of preference, and in adjusting current facility locations. A scaling factor $\alpha$ is used to mitigate the computational burden for the CPMP-M to find solutions. Constraint (9.2b) sets the number of polling facility $j$. Constraints (9.2c) stipulate that demand $i$ can be served by polling facility $j$ only if the facility is open (i.e., $x_j = 1$). Constraints (9.2d) and (9.2e) together allow the multiple assignment of demand $i$ to facility $j$ if more than two voting facilities $j$ are open. To indicate which facility $j$ has multiple polling stations from demand $i$, the decision variable $x_j$ is defined as positive integer value. The capacity of workload at facility $j$ should be limited by constraints (9.2f) to avoid overloading voting activity but run the minimum threshold of voting workload by constraints (9.2g). Constraints (9.2h) define assignment variables as binary.

## 9.5    Computational Experiments and Implications

The application was made to the study area, *Seogyo-Dong, Mapo-Gu*, Seoul, based on information from the 2016 election. In this area, 59 potential facilities were identified as candidates and 2497 demand areas were extracted at a household level from 21,378 eligible registered voters' address information. The processing tool in ArcGIS 10.4 was used for the geo-coding process and aggregation. In our numerical experiments, the known instance with current polling facilities was pre-solved as a reference. The capacity of facilities ranged from 2192 to 4297 based on data from the 2016 National Assembly election. The number of polling facilities was set to $p = 7$, as designated for this area by the POEA. The physical distance from demand areas to facilities was measured using ArcGIS 10.4. For comparison, several different facility capacity ranges from [2500, 4100] to [2500, 3100] by $-100$ from the upper bound of capacity were applied. Preference factor exponent β ranged from 0.0 to 3.0 by 0.5 in increments of 0.5. In terms of computational complexity, as is well known, the standard $p$-median problem is in the class of NP-hard, which indicates that the problems are solved with optimality to a limited size of instances [5, 38]. However, in our case study, all instances were solved to optimality using CPLEX 12.5 with an Intel Core$^{TM}$ i5-3570 running at 3.40 GHz with 4 GB RAM on the Windows 7 OS. Most instances were solved within a few minutes, but the worst instance took 15.2 min for the instance with β $= 0$ at the facility capacity range of [2500, 3800]. The reason for the solving capability of the CPMP-M was related to the relaxation of location decision variables $x_j$ from the definition as a binary value to positive real number. Table 9.1 presents the computational results of two cases with preference exponent β when $p = 7$. The two cases, the reference and the extreme case, were selected for comparison purposes. Note that a result of the instance with capacity [2192, 4297] when β $= 0$ was a reference and was the result of the current voting system of the study area as of the 2016 election. In this case, β $= 0$ was set to represent that no preference factor was involved in location-allocation decisions. In contrast, a greater β means greater influence of voter preference on selecting voting facilities over physical distance itself.

Three findings are worthy to note in drawing policy implications. First, objective function decreases with higher β and/or smaller upper facility capacity. The objective function may be influenced by what scaling factor is used; however, the tendency has been changed in our experiment. The result implies that the total utility

**Table 9.1** Computational results of the CPMP-M

| β | Capacity Lower | Upper | Objective value | Solution time (s) | APWD[a] (m) | Polling facility location |
|---|---|---|---|---|---|---|
| 0.0 | 2192 | 4297 | 3,758,429 | 120.4 | 176.5 | 16, 32, 33, 43, 47, 52, 56 |
| 0.5 | | | 3,210,506 | 165.1 | 183.8 | 6, 27, 32, 34, 44, 47, 52 |
| 1.0 | | | 2,656,564 | 192.8 | 184.8 | 6, 27, 32, 34, 44, 47, 52 |
| 1.5 | | | 2,146,974 | 55.8 | 203.3 | 6, 27, 32, 34, 39, 44, 47 |
| 2.0 | | | 1,721,321 | 117.9 | 216.6 | 6, 27, 32, 39, 44, 47, 53 |
| 2.5 | | | 1,362,473 | 180.8 | 252.0 | 6 (2)[b], 32, 39, 44, 52, 53 |
| 3.0 | | | 1,039,624 | 133.7 | 252.3 | 6 (2)[b], 32, 39, 44, 52, 53 |
| 0.0 | 2500 | 3500 | 3,828,205 | 697.9 | 179.8 | 16, 27, 34, 47, 49, 52, 58 |
| 0.5 | | | 3,288,007 | 464.9 | 187.6 | 6, 27, 32, 34, 42, 47, 52 |
| 1.0 | | | 2,708,279 | 156.6 | 200.0 | 6, 27, 32, 34, 39, 44, 47 |
| 1.5 | | | 2,186,976 | 51.7 | 202.6 | 6, 27, 32, 34, 39, 44, 47 |
| 2.0 | | | 1,775,225 | 123.3 | 205.8 | 6, 27, 32, 34, 39, 44, 47 |
| 2.5 | | | 1,407,920 | 219.9 | 236.4 | 6 (2)[b], 32, 34, 39, 44, 47 |
| 3.0 | | | 1,082,159 | 191.0 | 272.8 | 6 (3)[b], 32, 39, 44, 53 |

[a] APWD: Averaged population-weighted distance
[b] The facility houses multiple pulling stations

cost $(d_{ij}/f_j^\beta)$ of demands is improved. However, with an increase of β, meaning a greater preference for candidate facilities, a greater averaged population-weighted distance (APWD) is expected. This result, in a behavioral perspective of voters, indicates that voters will trade their increased travel cost for the searching cost spent to find new facilities using their pre-knowledge and experience. It should be highlighted that increased β results in the co-location of polling stations in a facility. For example, in the instances β = 2.5 and 3.0 for the capacity [2500, 3500], the numbers of selected facilities are 5 and 4, respectively. This is because a polling facility begins to house multiple polling stations (i.e., $x_6$ = 2 and 3) with higher β. This tendency is observed in the other instances within the same capacity having different β. In theory, if distance effect is a dominant factor, then the model tries to disperse facilities to cover the demand located near the facilities [22]. If distance factor is trivial, then the model tries to cover demands with a smaller number of facilities. In practice, managing a smaller number of voting facilities is beneficial, from an administrative perspective because of its ability to reduce operational costs. However, it entails a greater inconvenience for voters, because they need to travel farther to vote.

Second, the optimal arrangement of voting facilities and stations gives information on the ideal delineation among precincts. Figure 9.5 snapshots the results of two instances with different β(= 0, 3) under the identical capacity condition [2500, 3500]. The colored layers display the boundaries of the current seven precincts. Per the current requirements, voters in a precinct should go to one designated polling facility per precinct. However, the arrangement by the CPMP-M points out that the current location of polling facilities might not be selected to

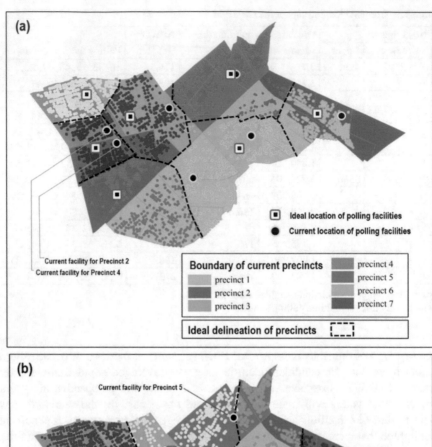

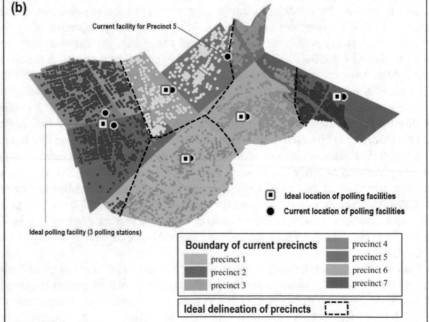

**Fig. 9.5** Spatial arrangement of polling facilities and ideal delineation of precincts. (**a**) β = 0, Capacity = [2500, 3500]. (**b**) β = 3, Capacity = [2500, 3500]

enhance voters' accessibility, raising the issue of adjusting the boundary of current precincts. In Fig. 9.5a, when only distance effect is considered, only one current location is re-selected by the model, and the other six current polling facilities are relocated. Interestingly, current polling facilities for precincts 2 and 4, which are very closely located to each other, can be replaced with a single facility $x_6$, which would require an adjustment of the boundary of both precincts. Similarly, current polling facilities of precincts 1 and 6 can be replaced with a new polling facility $x_{53}$. Based upon these replacements and the relocation of existing facilities, the ideal delineation of precincts for $p = 7$ is suggested (see dotted lines). The second map in Fig. 9.5b indicates another implication when $\beta = 3$. First, the locations of four current polling facilities overlap the location given by the model's result, implying that currently assigned facilities are based on voters' preferences. To make this arrangement of polling facilities valid, the model suggests that the current seven-precinct system can be replaced with a five-precinct system. Specifically, current precincts 2 and 4 can be merged into a single precinct, and new facility $x_6$ can take their place by housing three polling stations in the facility. The adjusted boundary of the model also presents all facilities centrally located within each precinct.

Third, in terms of facility capacity design, the sensitivity of capacity should be considered. Table 9.2 presents the relationships of the change of objective value and capacity ranges. With $\beta$ being equal, the objective value increases with a tighter facility capacity, especially to the maximum capacity. This result is well anticipated because a limited capacity of a facility can force voters to choose less congested facilities rather than go to nearer and/or preferable facilities. Of interest is the sensitivity to the different $\beta$. As presented in the last column, the capacity sensitivity is much higher for greater values of $\beta$. This result may imply that voters would respond more sensitively to the expected workload of the facility if the facility were designated on the basis of a historical familiarity. Voters' pre-knowledge of the facility's capacity may lead them to choose an alternative one. This tendency has been well observed in other cases from the 2016 election in South Korea. In some areas, a high percentage of turnout was found in small and local polling facilities (e.g., a set of polling stations within an apartment complex or parking space) rather than at single and large public facilities of a precinct, such as public schools and community centers.

**Table 9.2** Model response to the different capacity ranges of polling facility

| $\beta$ | Objective of current system capacity [2192–4297] | Objective of selective capacity ranges [2500–4100] | [2500–3900] | [2500–3700] | [2500–3500] | Capacity sensitivity[a] |
|---|---|---|---|---|---|---|
| 0.0 | 3,758,429 | 3,791,364 | 3,816,765 | 3,822,988 | 3,828,359 | 1.86% |
| 1.0 | 2,656,564 | 2,694,008 | 2,698,521 | 2,702,735 | 2,708,279 | 1.95% |
| 2.0 | 1,721,321 | 1,746,364 | 1,753,533 | 1,763,084 | 1,775,225 | 3.13% |
| 3.0 | 1,039,624 | 1,065,826 | 1,069,957 | 1,076,601 | 1,082,159 | 4.09% |

[a]Capacity sensitivity $= \dfrac{(Existing\ system's\ Obj.) - (Selective\ capacity\ range's\ Obj.)}{(Existing\ system's\ Obj.)}$

## 9.6 Concluding Remarks

Voting is a critical political activity that gives voters the opportunity and right to express their opinion in modern democratic society. However, whether voters participate in such political activity is not explained simply by distance effect or strong motivation as a citizen [7]. Rather, it is a mixture of psychological conditions and pre-knowledge based on the experience in selecting the best polling facility. Ways to increase voter turnout have been widely explored, as reviewed in the literature. However, the optimization approach is recognized as a crucial way of assessing the efficiency of the current system and drawing policy implications. This paper presents the CPMP-M, a location-allocation model for locating polling facilities and allocating stations to those facilities. Using the model helps policy-makers who are involved in the voting process to assess the existing system and find solutions that offer a more balanced workload of facilities along with improved accessibility. In a modeling structure, the CPMP-M stems from the traditional $p$-median problem; however, the model is formulated to be tailored to the specific requirements of location-allocation in the region of interest (i.e., South Korea). The model presents several important findings when deciding where to place polling facilities and stations. First, a traditional approach may take only the distance effect into account in the modeling process. However, voters' decisions are made based not only distance but also upon preference for facilities. The CPMP-M reflects both factors in an objective function, which is defined as "utility cost." If a voter's goal is to minimize the utility cost of finding a polling facility, his or her preference often strongly influences their choice. Our numerical results based on preference factors may induce the need for policy-makers to (a) relocate existing polling facilities, (b) merge certain precincts, and (c) adjust existing boundaries of precincts to enhance the efficiency of administration of the voting process. It may be an issue of how to quantify intangible factors, such as familiarity with existing facilities, pre-knowledge giving rising to search costs for finding the best place to vote, and perception of new facilities, into the model. However, the factor of voter preference is also emphasized, because high level of preference can entail the co-location of multiple ballot boxes in a single facility.

## References

1. Aldrich, J.H.: Rational choice and voter turnout. Am. J. Polit. Sci. **37**(1), 246–278 (1993)
2. Bhatti, Y.: Distance and voting: evidence from Danish municipalities. Scand. Polit. Stud. **35**(2), 141–158 (2012)
3. Brady, H.E., McNulty, J.E.: Turning out to vote: the costs of finding and getting to the polling place. Am. Polit. Sci. Rev. **105**(1), 115–134 (2011)
4. Cantoni, E.: A precinct too far: turnout and voting costs. Am. Econ. J. Appl. Econ. https://www.aeaweb.org/articles?id=10.1257/app.20180306&&from=f (last access: 5, December 2019)
5. Daskin, M.S., Maass K.L.: The $p$-median problem. In: Laporte, G., Nickel, S., Saldanha da Gamma, F. (eds.) Location Science, chap. 1, pp. 21–45. Springer, Cham (2015)

6. Downs, A.: An Economic Theory of Democracy. Harper and Row, New York (1957)
7. Dyck, J.J. Gimpel, J.G.: Distance, turnout, and the convenience of voting. Soc. Sci. Q. **86**(3), 531–548 (2005)
8. Fullmer, E.B.: Early voting: do more sites lead to higher turn-out? Election Law J. **14**(2), 81–96 (2015)
9. Ghiani, G., Guerriero, F., Musmanno, R.: The capacitated plant location problem with multiple facilities in the same site. Comput. Oper. Res. **29**(13), 1903–1912 (2002)
10. Ghoseiri, K., Ghannadpour, S.F.: Solving capacitated $p$-median problem using genetic algorithm. In: Helander, M., Xie, M., Jiao, R., Tan, K.C. (eds.) 2007 IEEE International Conference on Industrial Engineering and Engineering Management, Singapore, pp. 885–889 (2007)
11. Gimpel, J.G., Schuknecht, J.E.: Political participation and the accessibility of the ballot box. Polit. Geogr. **22**(5), 471–488 (2003)
12. Hanjoul, P., Peeters, D.: A facility location problem with clients' preference orderings. Reg. Sci. Urban Econ. **17**(3), 451–473 (1987)
13. Haspel, M., Knotts, H.G.: Location, location, location: precinct placement and the costs of voting. J. Polit. **67**(2), 560–573 (2005)
14. Hwang, A.R.: The effects of electoral district features on turnout: an analysis of the 15th Korean national assembly election. Korean Polit. Sci. Rev. **30**(4), 285–298 (1997)
15. Hwang, A.R.: Determinants of turnout rates in Korean local elections. J. Korean Assoc. Local Gov. Stud. **23**(1), 217–236 (2011)
16. Ingraham, C.: This is the Best Explanation of Gerrymandering You will Ever See. The Washington Post, Washington (2015). Available at https://www.washingtonpost.com/news/wonk/wp/2015/03/01/this-is-the-best-explanation-of-gerrymandering-you-will-ever-see/?noredirect=on&utm_term=.f6787372f547
17. Kang, H.W.: The criteria of redistricting and gerrymandering: a case study on the city of Pyongtack. Korean J. Polit. Sci. **12**(1), 321–342 (2004)
18. Kariv, O., Hakimi, S.L.: An algorithmic approach to network location problems. II: the $p$-medians. SIAM J. Appl. Math. **37**(3), 539–560 (1979)
19. Kim., W.: Institutional measures for increasing voter turnout. J. Contemp. Polit. **2**(1), 175–196 (2009)
20. Kim, M.J. Kim., K.: Spatial optimization approaches to redistricting for national assembly election: a case study on Yongin city. J. Korean Geogr. Soc. **48**(3), 387–401 (2013)
21. Kim, M.J., Kim., K.: National assembly redistricting corresponding to constitutional discordance adjudication: in the case of Gyeonggi. Geogr. J. Korea **50**(1), 1–13 (2016)
22. Kim, H., O'Kelly M.E.: Reliable $p$-hub location problems in telecommunication networks. Geogr. Anal. **41**(3), 283–306 (2009)
23. Kim, K., Lee, G., Shin, J.: A study on reconstructing of local administrative districts using spatial analysis and modeling. J. Korean Assoc. Reg. Geogr. **16**(6), 673–688 (2010)
24. Kim, H., Chun, Y., Kim, K.: Delimitation of functional regions using a $p$-regions problem approach. Int. Reg. Sci. Rev. **38**(3), 235–263 (2015)
25. Kim, K., Dean, D.J., Kim, H., Chun, Y.: Spatial optimization for regionalization problems with spatial interaction: a heuristic approach. Int. J. Geogr. Inf. Sci. **30**(3), 451–473 (2016)
26. Kim, K., Kim, H., Chun, Y.: A robust heuristic approach for political districting problems. In: Thill, J.-C., Dragicevic, S. (eds.) GeoComputational Analysis and Modeling of Regional Systems, pp. 305–324. Springer, Berlin (2018)
27. Konishi, K., Murata, T., Natori, R.: Voting simulation for improving voter turnout and reducing the number of polling places. J. Jpn Soc. Fuzzy Theory Intell. Inf. **22**(2), 203–210 (2010)
28. Lee, S.-I. Kim, K.: Representing the population density distribution of Seoul using dasymetric mapping techniques in a GIS environment. J. Korean Cartogr. Assoc. **7**(2), 53–67 (2007)
29. Lee, G., Kim, K.: Estimating de facto population using spatial statistics. J. Korean Cartogr. Assoc. **16**(2), 71–93 (2016)
30. Lee, S., Lee, S.W., Hong, B.Y., Eom, H., Shin, H.-S., Kim, K.-M.: Representation of population distribution based on residential building types by using the dasymetric mapping in Seoul. J. Korea Spat. Inf. Soc. **22**(3), 89–99 (2014)

31. Mansley, E., Demšar, U.: Space matters: geographic variability of electoral turnout determinants in the 2012 London mayoral election. Elect. Stud. **40**, 322–334 (2015)
32. McNulty, J.E., Dowling, C.M., Ariotti, M.H.: Driving saints to sin: how increasing the difficulty of voting dissuades even the most motivated voters. Polit. Anal. **17**(4), 435–455 (2009)
33. Murata, T., Konishi, K.: Making a practical policy proposal for polling place assignment using voting simulation tool. SICE J. Control, Meas. Sys. Integr. **6**(2), 124–130 (2013)
34. Murray, A.T., O'Kelly, M.E., Church, R.L.: Regional service coverage modeling. Comput. Oper. Res. **35**(2), 339–355 (2008)
35. Murray, A.T., Tong, D., Kim, K.: Enhancing classic coverage location models. Int. Reg. Sci. Rev. **33**(2), 115–133 (2010)
36. National Election Commission: Analysis of Turnout Rates of the 20th National Assembly Election (2016). Available at http://www.nec.go.kr/. Last Accessed 1 Jun 2018
37. Orford, S., Railings, C., Thrasher, M., Borisyuk, G.: Changes in the probability of voter turnout when resiting polling stations: a case study in Brent, UK. Eviron. Plann. C. Gov. Policy **29**(1), 149–169 (2011)
38. Reese, J.: Solution methods for the *p*-median problem: an annotated bibliography. Networks **48**(3), 125–142 (2006)
39. Rutchick, A.M.: Deus ex machina: the influence of polling place on voting behavior. Polit. Psychol. **31**(2), 209–225 (2010)
40. Stein, R.M., Vonnahme, G.: Effect of election day vote centers on voter participation. Election Law J. **11**(3), 291–301 (2012)
41. The Public Official Election Act: Statutes of the Republic of South Korea (2016). Available at https://elaw.klri.re.kr/. Last Accessed 18 Jun 2018
42. Tollison, R., Willet, T.: Some simple economics of voting and not voting. Public Choice **14**(1), 59–71 (1973)
43. Wang, X.J., Yang, M., Fry, M.J.: Efficiency and equity tradeoffs in voting machine allocation problems. J. Oper. Res. Soc. **66**(8), 1363–1369 (2015)
44. Williams, J.C.: Political redistricting: a review. Pap. Reg. Sci. **74**(1), 13–40 (1995)
45. Ye, H., Kim, H.: Measuring spatial health disparity using a network-based accessibility index method in a GIS environment: a case study of Hillsborough county, Florida. Int. J. Geospat. Environ. Res. **1**(1), Article 2 (2014)

# Chapter 10
# Territory Design for Sales Force Sizing

Juan G. Moya-García and M. Angélica Salazar-Aguilar

**Abstract** In sales territory design applications, a sales force team is in charge of performing recurring visits to the customers and typically, each territory is assigned to a sales representative with the aim to establish long-term personal relationship with the customers. At the strategic level, the decision maker must partition the set of customer in sales territories and at the tactical level, the daily routes (schedule of visits) of the sales representatives must be planned. Balanced sales territories allow better customer coverage and balanced workload. Additionally, efficient routes allow to perform more visits and to reduce the travel time. In this work, we focus in an application of territory design for determining the size of the sales force in a Mexican company. We also describe a simple heuristic for this problem and analyze its performance in two real cases from the company. Computational results show that the proposed heuristic produces high-quality solutions within a low computation time.

## 10.1 Introduction

The process to measure the productivity of sales force in some companies has been covered by several researches (see [2–4, 12, 15], among others). Frequently, companies evaluate the size of their sales force with the aim to determine the optimal human resources required to perform the activities involved in each sales channel (convenience stores, grocery stores, restaurants, indirect customers, etc.) in order

J. G. Moya-García
Linde México, Nuevo León, Mexico
e-mail: gerardo.moya@linde.com

M. A. Salazar-Aguilar (✉)
Graduate Program in Systems Engineering, Universidad Autónoma de Nuevo León (UANL), San Nicolás de los Garza, NL, Mexico
e-mail: maria.salazaragl@uanl.edu.mx

© Springer Nature Switzerland AG 2020
R. Z. Ríos-Mercado (ed.), *Optimal Districting and Territory Design*, International Series in Operations Research & Management Science 284,
https://doi.org/10.1007/978-3-030-34312-5_10

to ensure customer satisfaction in dynamic markets. There are several factors that make the company to review the size of its sales force periodically, such as:

1. Growth of the retail chains
2. Increase and complexity of in-store activities
3. Life time of some traditional outlets that face difficult times
4. Changes in the purchase patterns
5. Population growth and demographic issues.

Therefore, it is suggested to carry out a review of the sales structure every year and try to take advantage of the new opportunities present in the market.

In the literature, mathematical models have been developed in order to help companies to decide how large their sales force should be, see, for example, Lodish et al. [17]. An excellent overview of sales force deployment decisions models can also be found in Howick and Pidd [13] and in Salazar-Aguilar et al. [20].

A common approach to solve the sales force sizing problem consists in dividing the market in clusters, called territories, and determining the schedules of visits for the sales team on each territory. Besides, according to Kalcsics et al. [14], the main objective of territory design or districting process is to group geographic areas in clusters. So, shaping the territory for a sales person is known as sales territory design, as stated by Shanker et al. [21].

In this work, we present an overview of general applications of territory design and we focus in the ones related to sales territory design. Moreover, we describe a real situation faced by a nation-wide company in Mexico and propose a heuristic in which routing decisions and territory design are integrated to determine the size of the sales force.

The chapter is organized as follows. Section 10.2 provides a brief overview of related work. Section 10.3 describes a real sales force sizing problem. Section 10.3.1 presents a simple heuristic to solve the problem. Section 10.4 shows the performance of the proposed heuristic when solving two real cases of the sales force sizing problem. Finally, some conclusions are included in Sect. 10.5.

## 10.2   Related Work

Unbalanced sales territories may cause lower market share, slower growth, and workload unbalance to the companies. Therefore, the sales territory optimization is the key to increase sales performance, decrease the driving time, and have more manageable territories. There are early studies that propose integer programming formulations for territory design with the objective of maximizing compactness with workload balancing constraints, such as the work proposed by Hess and Samuels [11], where the number of desired territories is given and balance constraints depend on the number of customers, call duration, and visit frequencies.

Territory configurations can also be evaluated in terms of the total workload of a salesperson and the variation coefficient between the regions (see [7]). Other desired characteristics are the compactness and contiguity, which are present in the study of Fleischmann and Paraschis [8]. In the work of Ríos-Mercado and Salazar-Acosta [19], commercial territories are created by taking into account compactness, connectivity, customer balance, and demand criteria. Other models that consider multiple balancing criteria have been also presented in the research of Zoltners and Sinha [25]. A different, but also a common objective is to maximize the profit of each territory, this objective has been studied by Lodish [16], Zoltners [24], and Skiera and Albers [22].

As large-scale problems demand significant computational resources, in real implementations, solutions are approximated by heuristics. One example is the GRASP (Greedy Randomized Adaptive Search Procedure) proposed by Ríos-Mercado and Fernández [18], in which a commercial territory design problem with multiple balancing constraints is studied. In other cases, the territory design problems are addressed with hybrid procedures, see, for instance, the work of González-Ramírez et al. [9], which is a combination of GRASP and Tabu Search for solving a territory design problem for pickup and delivery operations for large-scale instances. As one can see, real applications not only include territory design decisions, they also could include, scheduling and routing decisions, such as the work proposed by Hervert-Escobar and Alexandrov [10], which combines the territory design, scheduling, and routing problems; these three problems individually have been shown to be NP-complete, therefore, the authors proposed a randomized block projection method for the territory design problem followed by a branch and bound algorithm for the scheduling and routing problems.

Besides territorial design, there are other related problems to sales force, see the work of Drexl and Haase [6], where the problem is divided in subproblems like sales force sizing, salesman location, territory sales alignment, and sales resource allocation; these problems are solved simultaneously, using a nonlinear mixed integer programming model.

In some types of territory design problems the workload is strongly related to the complexity of the distribution routes. Velarde Cantú et al. [23] propose a mixed integer linear programming formulation that integrates both territory design and routing decisions. The authors validate the mathematical model by solving a practical instance with 40 customers and 4 territories.

In this work we present a sales force sizing problem that includes decisions related to territory design and scheduling (routing) of visits. To the best of our knowledge, there are only two works closely related to our problem, the one proposed by Salazar-Aguilar et al. [20] and the one studied by Bender et al. [5]. In [20], the authors deal with the sales force sizing problem by solving a multi-period VRP with time windows and heterogeneous customers and vendors. They propose a mixed integer linear programming formulation that can be used to solve small instances of the problem, however, their work does not include territory design decisions. Besides, Bender et al. [5] propose the multi-period service territory design problem (MPSTDP-S), which consists of finding the visit schedule for each territory while compactness and service time balance between territories and working periods are satisfied. Two levels of clustering are considered: week clusters and day clusters; a location-allocation heuristic was proposed and tested on a large set of real instances; computational results showed the efficiency of their approach.

## 10.3  Sales Force Sizing Problem

The sales force sizing problem presented in this work arises from a goods distribution company that is interested in having more manageable sales territories and in decreasing the total costs associated to sales force and products distribution. The company is dedicated to produce and sale fast moving consumer goods and manages a portfolio of local and global brands. The finished product is sent from the factory to the distribution centers, depots, sub-depots, and cross-docking, and then to retailers that are classified in channels. For some of the channels the company uses customer's own supply network, while in others, the sales team is in charge of product distribution to all the customer locations. Hence, the problem we studied here can be described as follows.

Given a set of customers that must be visited every week and a depot where the sale representatives start and finish their journey each working day, the sales force sizing problem consists in determining the minimum number of vendors and the daily schedule of visits such that all customers are visited within a planning horizon $h$ (see [20]). Therefore, this problem can be seen as a sales territory design problem where the objective is to minimize the total number of territories needed to serve all customers, while the working time (service and travel times) on each territory does not exceed the time limit of a working week. In the territorial design context, the number of territories determines the size of the sales force, i.e., the number of sale representatives, and the daily schedules of visits that a sales person should perform during the week are determined by scheduling the set of customers included in the territory. Some side characteristics such as different kind of customers and skills of the sales representatives are considered.

Let $\Omega$ be the set of routes that start and finish at the depot and whose duration does not exceed the duration of a working day ($T_{max}$) and let $c(r)$ be the centroid of route $r \in \Omega$. Assume $a_{ir} = 1$ if route $r$ visits customer $i$. Then, a territory $X_k, k \in K$ will be composed by a set of routes such that the 1-median between them is minimized. Then, the centroid $c_k$ of a territory $X_k$ is computed by $c_k = \arg\min_{r \in X_k} \sum_{\bar{r} \in X_k} d_{c(r)c(\bar{r})}$, where $d_{c(r)c(\bar{r})}$ is the Euclidean distance between $c(r)$ and $c(\bar{r})$. Let $x_r^k$ be 1 if route $r$ is included in territory $k$, 0 otherwise; and let $y_k$ be 1 if territory $k$ is used, 0 otherwise. Then, a general formulation for the optimization problem can be written as follows:

$$\min \quad z = \lambda \sum_{k \in K} y_k + (1 - \lambda) \sum_{k \in K} \sum_{r \in \Omega} d_{c_k c(r)} x_r^k \tag{10.1a}$$

$$\text{subject to} \quad \sum_{k \in K} x_r^k = 1 \qquad\qquad r \in \Omega \tag{10.1b}$$

$$\sum_{k \in K} \sum_{r \in \Omega} a_{ir} x_r^k = 1 \qquad\qquad i \in N \tag{10.1c}$$

$$\sum_{r \in \Omega} x_r^k \le h y_k \qquad\qquad\qquad k \in K \qquad\qquad (10.1\text{d})$$

$$x_r^k \in \{0, 1\} \qquad\qquad\qquad k \in K,\, r \in \Omega \quad (10.1\text{e})$$

$$y_k \in \{0, 1\} \qquad\qquad\qquad k \in K \qquad\qquad (10.1\text{f})$$

Objective (10.1a) minimizes the total number of territories which determines the number of sales representatives meanwhile the dispersion of the territories is minimized. Constraints (10.1b) guarantee that each route is assigned to only one territory. Constraints (10.1c) assure that each customer is served by only one sales representative. Constraints (10.1d) establish the size of the territories, i.e., the number of routes included in each territory should be at most equal to the number of working days per week.

The implementation of a sales territorial design process besides to optimize the total number of territories has the following secondary objectives:

1. To reduce the gasoline expenses and vehicle maintenance through the reduction of the traveled distance within routes.
2. To increase productivity in terms of total visits, based on a routing and a compact territory design (in terms of distance).
3. To increase the number of visits with a sale by mean (called effective visits).
4. To avoid overtime in the sales team.

According to Albers et al. [1], sales managers are more suitable to work with heuristics rather than complex models. Indeed, they suggest to spend time investigating more simple heuristics to be easily understandable for practitioners and easy to implement. Therefore, due to the complexity of the problem, we propose a simple heuristic for the sales force sizing problem which is based on the decomposition of the problem into subproblems. The goal is to provide support to the decision making with respect to the size of the sales force and the scheduling of customer visits.

### 10.3.1 Routing-Clustering Heuristic

In this heuristic, basic units must be partitioned into sales territories which are desired to be compact (in terms of traveling and service times) and balanced (according to the workload of 6 working days). Then, two main stages are carried out: routes creation and routes clustering.

The preprocessing step consists of identifying the set of basic units that are in the instance on hand. In our study, the basic units are the basic geospatial areas defined by INEGI (National Institute of Geography and Statistics). Each basic unit has an associated service time and it encloses a set of individual customers of the same type (convenience stores, groceries, wholesalers, etc.). Traveling times between basic units are computed with a GIS (geographic information system).

Algorithm 1 displays the general steps of the proposed heuristic. For the routes creation, the algorithm starts with a partial route containing only one location, the central depot, then, the route is extended by adding the basic unit that is closest to the

---

**Algorithm 1** Routing-clustering heuristic

---

**Require:** Instance of the problem
**Ensure:** Sales territories

    // *Routing process*
1: Let $N_c$ be the set of basic units in channel $c$
2: **while** $N_c \neq \emptyset$ **do**
3:    Start with a partial route $r$ with just one basic unit $i$, which is the closest to the depot
4:    Let $r$ be the partial route and $k$ be the last basic unit in route $r$. Find basic unit $k'$ that is not yet in the routes and that is closer to $k$
5:    Insert $k'$ at the end of the partial route $r$
6:    **if** The total traveling time in the route plus the traveling time from $k'$ to the depot is less or equal than the time limit **then**
7:        $N_c \leftarrow N_c \setminus k'$
8:        go back to 4
9:    **else**
10:       remove $k'$ from the route and close the route $r$
11:       go back to 3
12:    **end if**
13: **end while**

    // *Local Search*
14: Let $R_c$ be the set of routes created in the routing process
15: **for** each route in $R_c$ **do**
16:    Explore consecutively the neighborhoods created by the moves: relocation, (1,1)-exchange, (2-0)-exchange, and (2-2)-exchange, until the first improvement is found or the full neighborhood has been explored.
17: **end for**
18: Destroy unproductive routes and reassign the basic units to other routes if possible.

    // *Clustering process*
19: **while** $R_c! = \emptyset$ **do**
20:    Start with a partial territory $T$ with just one route $r_i \in R_c$, whose centroid is the closest to the depot
21:    Let $T$ be the current partial territory. Find route $r_{k'}$ that is not yet in a territory and whose centroid is closer to route $r_k$.
22:    Insert route $r_{k'}$ at the end of the partial territory.
23:    $R_c \leftarrow R_c \setminus r_{k'}$
24:    **if** $|T| \leq 6$ or $R_c \neq \emptyset$ **then**
25:       go back to 21
26:    **else**
27:       close the current territory $T$
28:       STOP
29:    **end if**
30: **end while**

---

last unit assigned to the current route and whose traveling and service times do not exceed the available working time. This process is repeated while it exists a basic unit that can be added to the current route without exceeding the time limit (duration of a working day). If such a basic unit does not exist, the route is closed and another route is initialized. Notice that this routing process is iteratively repeated until all basic units belong to one route.

After having a set of feasible routes, four consecutive local searches are applied to each route in order to minimize their duration and then, to reduce the number of routes, if possible. The moves used in the local searches are the following:

- *Relocation*: this move sequentially selects one basic unit and reinserts it in another position in the route.
- *(1,1)-exchange*: this is known as a swap move and it consists of swapping the locations of two basic units.
- *(2-0)-exchange*: this move takes two consecutive basic units and relocate them in another position in the route.
- *(2-2)-exchange*: this is similar to (1-1)-exchange, the difference is that two chains of two consecutive basic units swap their locations in the route.

The local search process stops as soon as a better route is found (first improvement), then, a filter of routes takes place. In the real situation, the number of routes per channel should be multiple of six (due to the working days). Therefore, the most productive routes (multiple of six) per channel are kept and the unproductive routes (those with few basic units and long duration) are destroyed and the basic units assigned to them are re-inserted in other routes if it is possible.

The clustering process of routes consists of grouping sets of six routes each of which will conform a sales territory. The centroid of each route is computed as the average of the latitude and longitude of the basic units included in the route. Then, the distances between routes (centroids) are computed. A territory is initialized with the nearest centroid from the depot, and the five remaining routes are added by following a nearest neighbor heuristic (based on the centroids distance). This process is repeated until all routes are assigned to one territory. Notice that the number of territories determines the number of sales representatives to hire (one per territory).

## 10.4   Case Studies

The described heuristic has been implemented in Microsoft Excel and in order to evaluate its performance two real cases were studied: one in Mexico City and another in Monterrey. In the first case, a single channel of customers was evaluated. The second case was developed with the purpose of evaluating the algorithm in a depot with different customer channels and multiple types of profiles in the sales team. Notice that in both cases a weekly visit frequency and 6 working days per week were considered. The computational results are presented in the next subsections.

### 10.4.1  Mexico City Depot

This case study is composed by 1350 basic units (see Fig. 10.1) which include 10,183 active customers from December 2015. The territory design used by the company was composed by 38 territories.

After 73 min, 222 routes were created by our algorithm, then these routes were used to create the territories and 12 min later, the proposed algorithm reported a territory design with 37 territories, reaching a better compactness and ensuring minor travel distances in the routes than in the design implemented by the decision maker at the company. Figure 10.2 represents the map of the new territories and the graph of workload (in term of number of visits) per day, for each territory. One can notice that in most of the cases the balance between territories is reached.

Having the initial solution reported by our heuristic, we identified those routes with a few customers due to long travel times (see, for instance, Thursday's route of territory 6) and minor changes in territory boundaries were made in order to avoid crossing important streets. The total computation time to solve this case was 205 min which represents 7.1% of the time needed by the decision maker. Figure 10.3 shows this important time reduction in comparison with the traditional process, where an analyst in the company uses a geographic information system to group customers manually.

We observed that the change to this new territory design implies the reduction of one territory which saves about $20,000 USD per year. The proposal was accepted by the company and it was implemented at the beginning of 2016.

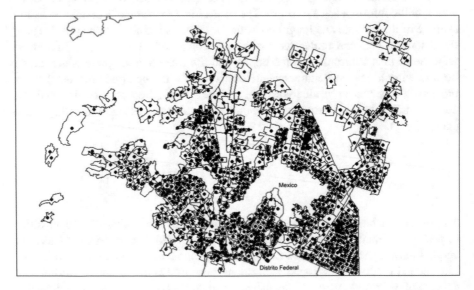

**Fig. 10.1** Distribution of basic units in the Mexico City depot

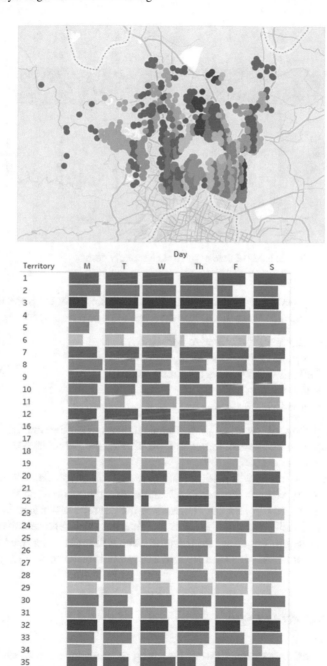

**Fig. 10.2** Map and route workload of territories in the Mexico City depot

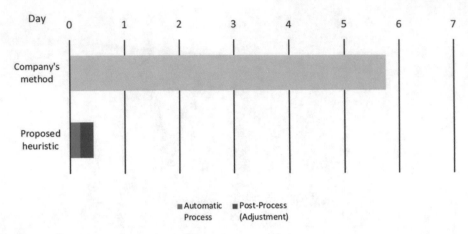

Fig. 10.3 Comparison in terms of time to carry out a territory design

To consistently evaluate the sales performance reached after the implementation, the following indicators were monitored on a monthly basis:

1. Average mileage per representative per week.
2. Number of average daily visits per representative.
3. Number of effective daily visits per representative.
4. Duration of the working day.
5. Average order size.

Figure 10.4 shows the average performance indicators for the first 6 months after the implementation and as a complement we present the percentage of change for each indicator in Table 10.1.

As one can see, the new territory design saves 17% of traveled kilometers and increases the daily visits in 2% with 14% more effective visits. The working time stills around 10 h per day and the decrease in the average order size is related to the increment of visits. Nevertheless, the fact that the sales representatives have to visit new zones requires training and time investment in order to get used to the new routes, which could contradicts the objective of not to impact the workday duration.

### 10.4.2   Monterrey Depot

In this case study the customers were classified in seven layers, according to their distribution channel and their service type, as we can see in Table 10.2. Then, the solution method was executed for each layer separately. Notice that the duration of a salesman visit varies according to the channel and service provided to the customer.

Solving this case study required 145 min and according to Fig. 10.5, the computation time increases with the number of customers (see layer 7).

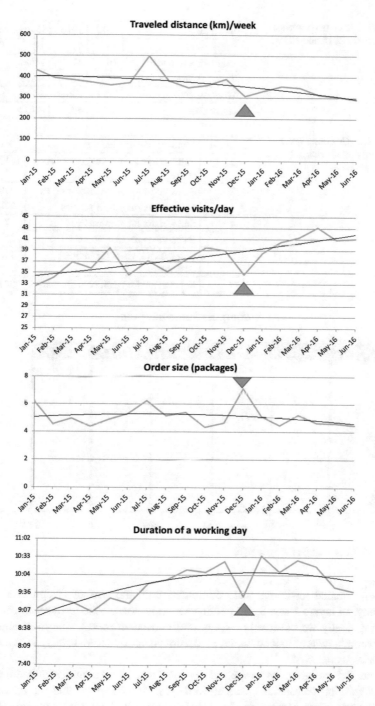

**Fig. 10.4** Performance trend from 2015 to the first 6 months after the implementation of the new sales territories

**Table 10.1** Key performance indicators for 2015 and 2016

| Indicator | 2015 | 2016 | % improvement |
|---|---|---|---|
| Daily visits | 50 | 51 | 2% |
| Daily effective visits | 36 | 41 | 14% |
| Traveled distance (km) | 388 | 324 | −17% |
| Working day | 9:39 | 10:09 | 5% |
| Order size | 5.3 | 4.8 | −10% |

**Table 10.2** Customer layers according to their channel and service type

| Layer | Channel and service type | Customers |
|---|---|---|
| 1 | Staff training visits for convenience stores | 1419 |
| 2 | Sales visits for convenience stores | 279 |
| 3 | Sales visits for grocery stores | 278 |
| 4 | Sales visits for wholesalers | 70 |
| 5 | Sales visits for restaurants | 137 |
| 6 | Stock check visits for indirect customers | 118 |
| 7 | Sales visits for traditional channel | 7931 |

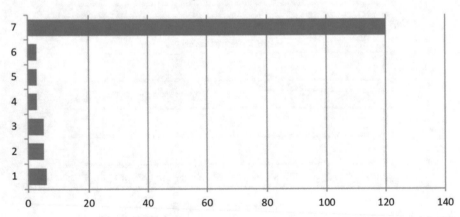

**Fig. 10.5** Computation time per layer

Then, created territories were compared with the current design used by the company. In layers 1 and 3, we obtained a reduction of one territory in each layer and graphically we observed compact and consistent routes. The most significant improvement was obtained in layer 2 (convenience store layer), in which two territories were saved.

Once the solution was reported by our heuristic, a minor adjustment was needed (similar to Mexico case), nevertheless, we observed that the balance and compactness were present in the solution generated by the heuristic. Besides, we noticed that in large instances it is possible to have poor routes in terms of

productivity, where most of the time is spend in driving. Figure 10.6 shows the
resultant territories and the visits balance chart for customers in layer 1.

At the end, the savings in terms of the number of territories was three. See
Table 10.3 for a complete comparison between the current solution used by the
company and the proposed design of territories obtained with our heuristic. In layers
1 and 3, the proposed solution saves one territory, and in layer 2, it saves two
territories. In contrast, in layer 7, the proposed solution increases one territory. In
summary, with the proposed territory design the company has about $60,000 USD
of potential savings.

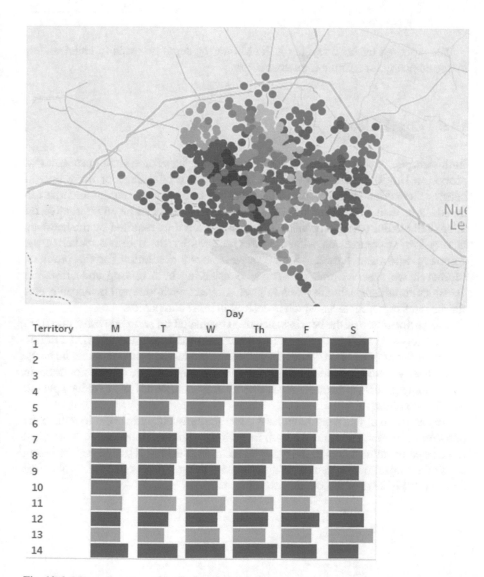

**Fig. 10.6** Map and route workload of territories in layer 1

**Table 10.3** Current and
proposed number of
territories per customer layer

| Layer | Current territories | Proposed territories |
|-------|---------------------|----------------------|
| 1     | 13                  | 12                   |
| 2     | 6                   | 4                    |
| 3     | 5                   | 4                    |
| 4     | 1                   | 1                    |
| 5     | 1                   | 1                    |
| 6     | 1                   | 1                    |
| 7     | 27                  | 28                   |
| Total | 54                  | 51                   |

The proposed territory design for the Monterrey depot is currently under review by the company for a future implementation.

## 10.5  Conclusions

In this chapter, we describe a territory design for a sales force sizing problem and propose a simple heuristic to solve it. An important advantage of the proposed heuristic is that it does not require a big investment in software or equipment, it can be coded in Visual Basic and run in MS Excel. Hence, it can be an alternative for micro and medium size companies. The computation time required by the heuristic is very low in comparison with the time required by the decision maker in the company, who takes between 5 or 6 days to create a solution for the problem. Moreover, we observed that the fact of considering both routing and clustering decisions simultaneously allows us to have compact territories and balanced routes for the sales representatives which is desired in most companies.

The performance of the heuristic has been evaluated on two real cases: one from Mexico City and another one from Monterrey. On each case, the reported solution decreases the number of territories, increases the productivity, and reduces the costs. Therefore, we encourage the managers to incorporate this type of decision tools in the planning process to increase service levels and at the same time to reduce overall distribution costs.

An extension of this work includes a general improvement in order to reduce the decision maker interventions. Indeed, in both case studies the decision maker had to make some adjustments to the reported solution. Besides, it would be interesting to see the impact on the quality of solutions by using network distances (obtained from a GIS) instead of the Euclidean distances.

# References

1. Albers, S., Raman, K., Lee, N.: Trends in optimization models of sales force management. J. Pers. Sell. Sales Manag. **35**(4), 275–291 (2015)
2. Anderson, E., Schmittlein, D.C.: Integration of the sales force: an empirical examination. Rand J. Econ. **15**(3), 385–395 (1984)
3. Avlonitis, G.J., Panagopoulos, N.G.: Selling and sales management: an introduction to the special section and recommendations on advancing the sales research agenda. Ind. Mark. Manag. **39**(7), 1045–1048 (2010)
4. Babakus, E., Cravens, D.W., Grant, K., Ingram, T.N., LaForge, R.W.: Investigating the relationships among sales, management control, sales territory design, salesperson performance, and sales organization effectiveness. Int. J. Res. Mark. **13**(4), 345–363 (1996)
5. Bender, M., Meyer, A., Kalcsics, J., Nickel, S.: The multi-period service territory design problem—an introduction, a model and a heuristic approach. Transp. Res. E: Logist. Transp. Rev. **96**, 135–157 (2016)
6. Drexl, A., Haase, K.: Fast approximation methods for sales force deployment. Manag. Sci. **45**(10), 1307–1323 (1999)
7. Easingwood, C.: A heuristic approach to selecting sales regions and territories. Oper. Res. Quart. **24**(4), 527–534 (1973)
8. Fleischmann, B., Paraschis, J.N.: Solving a large scale districting problem: a case report. Comput. Oper. Res. **15**(6), 521–533 (1988)
9. González-Ramírez, R.G., Smith, N.R., Askin, R.G., Camacho-Vallejo, J.F., González-Velarde, J.L.: A GRASP-Tabu heuristic approach to territory design for pickup and delivery operations for large-scale instances. Math. Prob. Eng. **2017**, 4708135 (2017)
10. Hervert-Escobar, L., Alexandrov, V.: Iterative projection approach for solving the territorial business sales optimization problem. Proc. Comput. Sci. **122**, 1069–1076 (2017)
11. Hess, S.W., Samuels, S.A.: Experiences with a sales districting model: criteria and implementation. Manag. Sci. **18**(4), 998–1006 (1971)
12. Horsky, D., Nelson, P.: Evaluation of salesforce size and productivity through efficient frontier benchmarking. Market. Sci. **15**(4), 301–320 (1996)
13. Howick, R., Pidd, M.: Sales force deployment models. Eur. J. Oper. Res. **48**(3), 295–310 (1990)
14. Kalcsics, J., Nickel, S., Schröder, M.: Towards a unified territorial design approach—applications, algorithms and GIS integration. TOP **13**(1), 1–56 (2005)
15. Ledingham, D., Kovac, M., Simon, H.L.: The new science of sales force productivity. Harv. Bus. Rev. **84**(9), 124–133 (2006)
16. Lodish, L.M.: Sales territory alignment to maximize profit. J. Market. Res. **12**(1), 30–36 (1975)
17. Lodish, L.M., Curtis, E., Ness, M., Simpson, M.K.: Sales force sizing and deployment using a decision calculus model at Syntex Laboratories. Interfaces **18**(1), 5–20 (1988)
18. Ríos-Mercado, R.Z., Fernández, E.: A reactive GRASP for a commercial territory design problem with multiple balancing requirements. Comput. Oper. Res. **36**(3), 755–776 (2009)
19. Ríos-Mercado, R.Z., Salazar-Acosta, J.C.: A GRASP with strategic oscillation for a commercial territory design problem with a routing budget constraint. In: Batyrshin, I., Sidorov, G. (eds.) Advances in Soft Computing. Lecture Notes in Artificial Intelligence, vol. 7095, pp. 307–318. Springer, Heidelberg (2011)
20. Salazar-Aguilar, M.A., Boyer, V., Sanchez Nigenda, R., Martínez-Salazar, I.A.: The sales force sizing problem with multi-period workload assignments, and service time windows. Cent. Eur. J. Oper. Res. **27**(1), 199–218 (2019)
21. Shanker, R.J., Turner, R.E., Zoltners, A.A.: Sales territory design: an integrated approach. Manag. Sci. **22**(3), 309–320 (1975)
22. Skiera, B., Albers, S.: Costa: contribution optimizing sales territory alignment. Market. Sci. **17**(3), 196–213 (1998)

23. Velarde Cantú, J.M., Bueno Solano, A., Lagarda Leyva, E.A., Lopez Acosta, M.: Optimization of territories and transport routes for hazardous products in a distribution network. J. Ind. Eng. Manag. **10**(4), 604–622 (2017)
24. Zoltners, A.A.: Integer programming models for sales territory alignment to maximize profit. J. Market. Res. **13**(4), 426–430 (1976)
25. Zoltners, A.A., Sinha, P.: Sales territory alignment: a review and model. Manag. Sci. **29**(11), 1237–1256 (1983)

# Index

**A**

Ambulance dispatching, 47, 156–157
Ambulance location, 47, 48, 156, 165
Ambulance redeployment, 155
Annotated bibliography, 18–26
   *See also* Police districting problem
Assignment problem (AP), 88–89
Averaged population-weighted distance
   (APWD), 185

**B**

Backup coverage, 13, 16, 22, 47
Balanced police sectors, 24
Balanced workload, 7, 11
Basic units (BUs), 80–81
Beardwood–Halton–Hammersley Theorem, 58
Big Data and Machine Learning, 52
Borsuk–Ulam theorem, 67
Branch-and-price algorithm, 6, 79, 133, 144,
   146–151
Brouwer's fixed-point theorem, 65
Buffalo Police Department, 16, 19

**C**

Cake cutting, 71–72
Capacitated p-Median Problem with
   Multiple Stations in the Same Facility
   (CPMP-M), 174, 182–185, 188
Capacity-demand match, 41, 44

Central District of Madrid, 24
Chilean National Police Force, 24
Classical operational research models, 25
Co-extensiveness, 35, 42, 44
Compactness, 3, 6
Computational geometry
   algorithms, 57–58
   geometric probability theory, 58
   notational conventions, 58
   shape constraints, 58
   Voronoi partition, 59
Computational results
   CPU time, 96–101
   Lagrangian relaxation approach, 92
   lower bounds, 93–96
   quadratic programming formulation, 92
   solutions, 102
   territory design problem, 92
Computer-aided dispatch (CAD) system, 154
Contiguity, 3–5
Contiguity criterion, 106, 114
Continuity of care, 35, 40
Convex partition, 69
Cooper's location-allocation method, 144
Crime heat-map, 11–12

**D**

Decision support system, 25
Designing police patrol sectors, 18
Discrete-event simulation, 6, 14, 23, 164, 166

© Springer Nature Switzerland AG 2020
R. Z. Ríos-Mercado (ed.), *Optimal Districting and Territory Design*, International
Series in Operations Research & Management Science 284,
https://doi.org/10.1007/978-3-030-34312-5

Printed in the United States
By Bookmasters